Today's Curiosity is Tomorrow's Cure

Today's Curiosity is Tomorrow's Cure

The Case for Basic Biomedical Research

Steve Caplan, PhD
Professor of Biochemistry and Molecular Biology
University of Nebraska Medical Center

CRC Press is an imprint of the
Taylor & Francis Group, an **informa** business

First edition published 2022
by CRC Press
6000 Broken Sound Parkway NW, Suite 300, Boca Raton, FL 33487-2742

and by CRC Press
2 Park Square, Milton Park, Abingdon, Oxon, OX14 4RN

CRC Press is an imprint of Taylor & Francis Group, LLC

Library of Congress Cataloging–in–Publication Data

Names: Caplan, Steve, author.
Title: Today's curiosity is tomorrow's cure : the case for basic biomedical research /
Steve Caplan, PhD, Professor of Biochemistry and Molecular Biology,
University of Nebraska Medical Center.
Description: First edition. | Boca Raton : CRC Press, 2022. |
Includes bibliographical references and index.
Identifiers: LCCN 2021021786 | ISBN 9781032065083 (paperback) |
ISBN 9781032065953 (hardback) | ISBN 9781003202974 (ebook)
Subjects: LCSH: Medical sciences--Research.
Classification: LCC R850 .C37 2022 | DDC 610.72--dc23
LC record available at https://lccn.loc.gov/2021021786

ISBN: 978-1-032-06595-3 (hbk)
ISBN: 978-1-032-06508-3 (pbk)
ISBN: 978-1-003-20297-4 (ebk)

DOI: 10.1201/9781003202974

Typeset in Times
by Deanta Global Publishing Services, Chennai, India

This book is dedicated to all types of scientists throughout history. To human curiosity and perseverance, and all those who strive to make sense of our world. From the teachers of science in schools, to those who train the next generation of scientists at undergraduate institutions, to those engaged in active research in the laboratory: no finding is too small, insignificant, or unworthy, as long as valid scientific methods are maintained. From the most prestigious Nobel laureates to the earliest stage undergraduate researchers, every scientist is a part of the vast fabric that advances and improves the human condition. No one can predict where the next big breakthrough will come from, but one prediction can accurately be made: the greater the investment in science and the more scientific research conducted, the more breakthroughs there will be. Scientists, take pride!

Contents

---

* Non-scientists should be able to understand and appreciate the majority of this chapter.

** Some of the details in this chapter will require some scientific background.

*** While the general principles in this chapter are understandable to the non-specialist reader, the technical and scientific details require a strong scientific background and may be more challenging for a layperson.

Acknowledgments

I would like to express my sincere gratitude to my immediate family, who as scientists and members of the scientific community have given me support and encouragement throughout the research and writing process. In addition, they have put up with my endless succession of ramblings about the science, history, experiments and personalities connected to the great biomedical discoveries, and all with well-intentioned good humor. I would like to express my special appreciation for the support and artistic skills of my wife, Dr. Naava Naslavsky, who was instrumental in bringing to life the many schematic diagrams and illustrations that are included within. In addition, I thank my friend Cindy Opler for kindly and bravely offering to read chapters as my "guinea-pig layperson," and providing me with advice. Finally, I am deeply grateful to my canine soul mate, Ginger the Vizsla-Labrador Retriever, who for eight terrific years following her "rescue" from the shelter served as my constant companion and patiently endured the ups and downs of the writing process with me. She was my "north-star" and our memories of her will always be treasured.

Preface

I am a scientist, one who has been performing independent, basic research in the confines of my own laboratory for the past 18 years. As an undergraduate student, I first became fascinated with receptors, basic biochemistry, and the field known as *signal transduction*; essentially how receptors on the surface of cells transmit various signals to the cell interior to instruct the cell ultimately to prepare for growth, division, migration, and any of the myriad functions carried out by cells. When I began my research toward a master's degree, I became drawn to the fields of immunology and cancer, with a desire to do something more practical, or as the term has become known in twenty-first-century science, *translational* (as in *translating* discoveries from the bench to the bedside). My research was focused for several years on macrophages, which are key cells involved in the first-line defense of the immune system against bacteria and other pathogens. However, even back in the 1990s it was suspected that they were also involved in *immune surveillance*, a term used to describe the early detection and destruction of individual cancer cells before they have an opportunity to expand into a full-blown tumor. My project involved the identification and characterization of a molecule that specifically activated these macrophages for "seek and destroy" missions, to make them effective at killing tumor cells (Caplan, Gallily, and Barenholz 1994).

In the course of my master's research, as a student at the Hebrew University in Jerusalem, Israel, at the Hadassah Medical Center at Ein Kerem, I had the opportunity to attend a series of lectures given by Dr. J. Michael Bishop, the eminent virologist and immunologist who received the 1989 Nobel Prize in Physiology or Medicine. It was perhaps Dr. Bishop, and a strong plug that he made for basic, curiosity-driven research, that led me to revise my thinking and realign my career goals. Accordingly, I chose to do a PhD in immunology, working on T cells and receptors and signal transduction (back to my original undergraduate interests), on one of the most complex receptors in a crucial immune system cell (Caplan, Almogi-Hazan, et al. 2001, Caplan and Baniyash 1995, 1996, 2000, Caplan et al. 1995).

When I completed my PhD in 1998, I moved to the National Institutes of Health in Bethesda, Maryland. Despite my intense interest in immunology, during my PhD stint I fell hopelessly in love with what is known loosely as "basic cell biology." This is an area that encompasses all things at the level of the cell and smaller, including the mechanisms of cell activation, division, migration, and every other basic cellular function that cells carry out in the course of their existence. Incumbent within this area of specialization, along with the classic biochemistry and biochemical methods used for such research, was an area with which I had little experience: light microscopy. Indeed, microscopy plays an increasingly powerful role in biomedical research by enabling scientists to visualize structures and organelles deep within the cell. Advances in light microscopy have led to a revolution in enhanced resolution, facilitating the imaging of organelles and even individual proteins within living cells. The major breakthroughs in microscopy techniques coupled with significant advances in quantifying and evaluating microscopy data have led to light microscopy becoming a highly sought after and incredibly versatile realm of expertise. I was determined to join this revolution.

Initially, joining the revolution was not an easy task for me; my first attempts to observe proteins tagged with a jellyfish-derived green fluorescent protein (to make the proteins visible) in a dark room under an epifluorescence microscope were a stark reminder that I was a "landlubber" who could get seasick in a rocking chair. Rather jarringly, within seconds of searching for the green-light-emitting proteins among the dark fields of cells, I could feel waves of nausea seeping in, and my first thought was that "this cell biology isn't for me." Fortunately, I got my "microscope legs" over time, and overcame my predisposition for motion sickness—at least in microscopy. But I still get seasick in a bathtub, and to this day (barring my naturalist's dream trip to the Darwin's Galapagos Islands in 1991), I do my best to stay on dry land.

My postdoctoral research was marked by attempts to study proteins involved in the regulation of endocytic pathways—to understand the fate of receptors on the cell surface that have bound to an external ligand or protein and have thus been *internalized* and admitted into the cell interior. Outstanding questions of the type that I addressed (and continue to address) are which proteins control the fate of these internalized receptors, and how they do so—for example how are they *sorted*? In other words, what controls whether they are returned back to the cell membrane to function again in binding new ligand (recycled), or whether they are destined to be taken apart inside the cell (degraded) so that their components, the individual amino acids which are the building blocks of proteins, can be reused in the assembly of newly synthesized proteins in the cell (Caplan et al. 2000, Caplan, Hartnell et al. 2001, Caplan et al. 2002, Dell'Angelica et al. 2000).

Having established my own independent research laboratory in 2003, I have spent the last two decades delighting in the freedom of being able to address significant basic research questions of my choosing. I am continually stimulated by the serendipitous research findings that my group has been making over the years—some of these findings have led so far outside of my expertise and comfort zone that I have required external expertise and have been very fortunate to have wonderful collaborations with many colleagues across the globe. But despite constant pressure to be drawn toward *translational science*, to carry out direct disease-related research, my core research has remained nonetheless truly basic (Caplan 2012).

Dr. Jon Lorsch, director of the National Institute of General Medical Sciences (NIGMS), noted in an interview with *Scientia* (*Scientia* 2017):

> History indicates that letting scientists "follow their noses" is the most productive path to medical and technological breakthroughs. Targeted projects focused on a particular application can be vitally important, for instance, the development of a vaccine for an emerging viral disease. But investigator-initiated research, which involves a combination of curiosity, expertise, creativity, and serendipity, has repeatedly proven to yield discoveries that are the most broadly applicable and transformative. By emphasizing investigator-initiated fundamental research, NIGMS is unleashing the creativity and energy of scientists across the country.

As a researcher who has been fortunate enough to derive my support primarily from NIGMS, I could not agree more!

Introduction

Today's Curiosity is Tomorrow's Cure: The Case for Basic Biomedical Science is not meant to be a comprehensive history of all of the great discoveries in biomedical research over the last ~150 years; indeed, that would be a near-impossible task. Instead, this book is aimed at highlighting a selection of key discoveries, and detailing the contextual background of each discovery, with special emphasis on demonstrating that basic science ultimately has a significant impact on improving health. In addition, the examples chosen in the chapters of this book also represent my own subjective choices, which are heavily influenced by the realms of my own research interests. Accordingly, this book contains many chapters discussing great discoveries focusing on biochemistry and molecular cell biology. On the one hand I chose to detail the discovery of DNA from its infancy, to our understanding of its involvement in heredity, the cracking of the genetic code, and the incredible advances in genetic engineering and genome editing that bring the promise of new treatments in the twenty-first century. On the other hand, I also highlight discoveries of some of the key cellular organelles (cilia, lysosomes, mitochondria, the Golgi complex, etc.) and how understanding their biogenesis and/or function has similarly led to breakthroughs in treating various diseases. One especially relevant point and a recurrent theme throughout the book is that rarely are discoveries a single *Eureka moment*; indeed, for the most part, great discoveries come from years or even decades of research, often from laboratories around the world investigating different, parallel, or overlapping research topics. Sometimes important discoveries are entirely serendipitous, while other times they result from targeted research. In some cases, it was necessary to await the development of new technologies (in physics, optics, biophysics, and chemistry, for example) so that the biomedical discoveries could be made. One thing has become clear in recent years, however: in the twenty-first century each important new discovery relies heavily on those made decades earlier, just as the roof of a house cannot be built unless the walls and foundation are first put in place. Indeed, today's new advances are increasingly reliant upon multiple findings from the past, as science and scientific research coalesce around a fundamental understanding of the workings of the cell.

While it is impossible to write a science book that will accommodate readers of every level of understanding, *Why Today's Curiosity is Tomorrow's Cure: The Case for Basic Biomedical Science* is written in a manner that should allow most laypersons a generalized grasp of the basic points relating to the mode of scientific discovery. At the same time, there is sufficient scientific, historical, and contextual detail to be instructive to undergraduate and graduate biology students. Accordingly, the book is a hybrid that intends to inform the general public, but also to provide actual scientific lessons for those who are interested in understanding the basics of the science and how it evolved. In particular, the many schematic diagrams are designed to provide clear but concise scientific explanations of the principles of the discoveries. In addition, each chapter has been evaluated for its accessibility to the general public: a rating of one asterisk indicates that a layperson or non-scientist should be able to understand and appreciate the vast majority of that chapter, a rating of two asterisks suggests that while a layperson should be able to understand the majority of principles and key points outlined in the chapter, there may be a small amount of scientific detail that requires more effort or some additional background. Finally, a rating of three asterisks indicates that the chapter is designed so that the general principles should be understandable to a non-specialist audience, but it does include a little more technical and scientific detail so that some of the finer points may be a little more challenging for a layperson. Nonetheless, every chapter, including those with three asterisks, should still be accessible to the general public so that a layperson understands the general concepts, history, and context behind the great discoveries described.

This introduction to the book has been updated and reupdated multiple times in the course of the 2020 pandemic, as science has been continually attacked and scientific researchers, experts, and infectious disease doctors are being challenged by ignorance. The internet, notwithstanding its significant contributions to the advancement of science, has at the same time been perilous for science advocacy. The inability of internet subscribers to differentiate between experts on a given subject and those charlatans who merely pretend to be experts has wreaked havoc on the public's perception of science. In a sense, the internet has essentially "leveled the playing field" by giving an almost equal legitimacy to every individual's own viewpoint, whether or not it is an expert's knowledge or a vantage point entirely devoid of knowledge—or even a conspiracy theory. Indeed, it will take many years to build a strong enough foundation of critical thinking to reinvigorate the general population and provide it not just with a knowledge of science and its workings, but also with the tools to discern what scientific expertise is, and who has legitimately acquired it. In part, this book aims to be a part of the long journey toward getting to that point.

The 2020 coronavirus pandemic, beyond the immense effect it has already had on the world as we know it, has further highlighted the need for this book. The 2016–2020 US administration, when initially faced with an urgent need for funding to combat the pandemic, immediately decided to reallocate resources previously awarded for Ebola research to the current crisis, emphasizing the short-sightedness of politicians who move from putting out one fire to putting out the next, rather than investing in prevention in the first place. It is precisely because of our understanding of the basic nature of viruses and human cells, spurred on by basic curiosity-driven research, that there is sufficient expertise available today to allow us to strategize and move rapidly toward identifying helpful drugs and implementing new clinical trials. It is our knowledge of how viruses make copies of themselves, how their RNA and DNA are replicated and transcribed, as well as their need for proteases in the replication pathway, that has spurred researchers and doctors to propose the use of drugs such as Remdesivir for clinical trials in infected patients. Moreover, an understanding of the basic mechanisms of DNA replication, transcription, and protein translation, essential processes that must occur in cells for division and normal function, has led to quiet but crucial breakthroughs in vaccine design. Indeed, rather than having to grow large quantities of virus, an expensive and time-consuming process, or alternatively, having to engineer, generate, and purify large amounts of viral protein to use as an immunogen for vaccines, researchers are now synthesizing the genetic viral material that codes for the viral protein, messenger RNA (mRNA), and introducing it directly into the body. Once taken up by human cells, the artificially synthesized viral mRNA is converted into viral protein in the body's cells to elicit an immune response and provide protection against the virus. The first of the two mRNA vaccines (Pfizer BioNTech BNT162b2 vaccine) designed to inoculate the population against COVID-19 proved to be over 94% effective in Phase 3 trials in a massive trial in the state of Israel (Dagan et al. 2021), while its Moderna counterpart mRNA-1273 vaccine exhibited about 95% efficacy. The Pfizer BioNTech vaccine has recently been fully approved by the US Federal Drug Administration (FDA) after an initial emergency authorization process, while the Moderna vaccine has emergency authorization and has filed recently for full approval. In the US alone, as of this writing, more than 209 million people have received at least one dose of the vaccine with 179 million people fully vaccinated (mostly with the mRNA vaccines), thus far representing nearly 64% of the eligible population above the age of 12. It is often easy to forget that without a basic understanding of DNA, without cracking the genetic code and understanding how proteins are translated in the cell—scientific findings that are only 50–60 years old and becoming better understood even now—these novel technologies for the generation of vaccines would not even be on the horizon.

Thus, the COVID-19 pandemic further highlights the need for basic research and unequivocally demonstrates the enormous value of basic research even in cases where the casual observer might think that the advances are purely from more clinical-based research. Another such example stems from the history of the discovery of insulin. Science historian Michael Bliss, in his book detailing the history of the discovery of insulin, describes how researchers set out to identify the factor that regulates glucose/carbohydrate metabolism (Bliss 1982). While that goal may appear to be very clinically oriented, it is clear from Bliss' depiction that major advances were limited until basic scientists managed to work out accurate methods to measure blood glucose levels, again reiterating how the advances in basic science and technology are absolutely crucial so that clinical breakthroughs can happen.

Upon entering the twenty-first century, for the past 20 years or so, there have been increasing calls for biomedical scientists to focus on *disease-related research.* As noted, such research is commonly known now as *translational research* with the notion that the findings derived from such research are translated to treatments and cures—to cure cancer and other diseases. There is significant pressure on researchers to spend less time following basic and curiosity-driven research, but rather to use government funding for more "relevant" disease-related research. The US Congress has taken an increasing role in these calls without a firm understanding of the significance of basic research, and in some cases has even asked to oversee and ensure that federally funded research dollars are used exclusively for researching and targeting specific diseases. Indeed, there have been an increasing and alarming number of reports that individual peer-reviewed grants that have been awarded to scientists are being scrutinized by some members of Congress who have little understanding of the meaning and value of basic biomedical research (Philippidis). A dozen years ago, a prominent politician and vice-presidential candidate actually derided researchers for spending US taxpayer dollars on fruit fly research, not realizing the significance of this organism as a key genetic model to spur novel discoveries that impact health human. Just recently, 77 Nobel laureates wrote to the Secretary of Health and Human Services demanding an investigation into the government's decision to intervene and negate the National Institutes of Health's (NIH's) expert grant review process and withdraw funding from an investigator (Science News Staff 2020).

The goal of this book is not to attack translational, pre-clinical, and clinical research, all of which are not only extremely important avenues of research, but ultimately the long-term goal of all biomedical research. Rather, the purpose of this book is to highlight that without continued and widespread investment in *basic* research, where both targeted and serendipitous discoveries have traditionally led to major advances in improving human health, there will be very few new translational and clinical research breakthroughs in the next 10–20 years. Basic research is the very pipeline that feeds all of the treatments and cures that are being tested today. Nobel laureate Arthur Kornberg once said: "No matter how counterintuitive it may seem, basic research has proven over and over to be the lifeline of practical advances in medicine. Without advances, medicine regresses and reverts to witchcraft" (Kornberg 1959b).

Why is basic science so important? A simple example is as follows: today new drugs are being tested in cells and in animal models to study their impact on infectious and non-infectious diseases. This is translational research. If such translational research shows promise, clinical studies may be designed to test the safety and efficacy of these promising drugs in clinical trials. But where do the new drugs come from?

For new drugs to be discovered, researchers need to understand the workings of cells. It is necessary to decipher what the ~35,000 known individual proteins do within the cell, how they are targeted to their correct locations in the cell, and how they cooperate with one another to carry out their complex functions. In addition, understanding the functions of those proteins, what activates them, turns them "on" and "off" and slows them down, and understanding their complex roles in the cell—all of this knowledge helps researchers elucidate what cellular targets are worthy of attention, thus allowing scientists new ways to modulate cellular function or dysfunction. A more complete understanding of these systems leads to new opportunities for drug interventions to be tested.

The NIH, the largest federal grant agency for biomedical research in the US, is comprised of 27 different institutes and centers, most of which are oriented toward select diseases. For example, the National Heart Lung and Blood Institute (NHLBI) provides grant support competitively to researchers who study diseases related to cardiovascular and pulmonary function. The National Institute for Diabetes and Digestive and Kidney Diseases (NIDDK) obviously promotes and supports research related to diabetes and renal diseases. NCATS is the National Center for Advancing Translational Sciences. Of the NIH institutes, only a single institute, the National Institute for General Medical Sciences (NIGMS) is designated to support the most basic biomedical science. It is important to stress that basic researchers *do* see their research as disease-oriented; however, unlike the research supported by the other NIH institutes, NIGMS-sponsored research often does not necessarily specify or know in advance *which* disease will benefit from the research. Such is the nature of basic biomedical research.

Whether science is purely basic, or whether it entails translational and/or clinical aspects, it is important to stress that no matter how significant a breakthrough may seem, and however innovative and remarkable a new biomedical advance may be, all science is built on a firm foundation of previous findings, many of which come from basic science. Although many great scientific leaps and opportunistic findings "favor the prepared mind" as once reportedly stated by one of the great microbiologists of the nineteenth century, Louis Pasteur, the significance of what a *prepared mind* means is often ignored. A prepared mind requires recognition and knowledge of prior research and findings, so that researchers are fully cognizant of what is known and what remains unknown, allowing them to ask the right questions and advance their scientific understanding beyond what is already known.

Indeed, biomedical breakthroughs often come as a culmination of decade-long studies from researchers around the globe, frequently spurred on by intense competition, willing collaboration, and the sharing of results, even prior to publication. As an example, Gross and Sepkowitz illustrate the immense journey from Jenner and his work on cowpox on to the eventual smallpox vaccine that finally eradicated this disease (Gross and Sepkowitz 1998). Almost 200 years passed from the time Jenner published his work "On the Origin of Vaccine Inoculation" in 1801 describing his experiments inoculating people with cowpox sores to prevent the contraction of smallpox (Jenner 1801), until 1980 when the World Health Organization finally declared that the world had been rid of this horrid disease.

Why is it so difficult for the general public to understand that breakthroughs in "useful science" and medicine are often the culmination of decades of basic research? Why can politicians and policy makers not grasp the idea that curiosity-driven research ultimately leads to the greatest discoveries? In her book *Science Unshackled*, C. Renee James contends that in a society where people complain that microwave ovens are too slow in heating up food, and where people are unwilling to wait 30 seconds to download internet files, instant gratification is the norm (James 2014). With the glacial pace at which basic science ultimately translates into useful biomedical treatments and cures, over years, decades, and sometimes even centuries, it is no wonder that people are not willing to wait.

Although this book is aimed at furthering the understanding of how basic research has led to key biomedical breakthroughs in the twenty-first and twentieth centuries, any attempt to discuss advances in basic science cannot be seriously considered without turning first to the truly earth-shattering observations and conclusions drawn by Charles Darwin, as outlined in *On the Origin of Species* published in 1859 (Darwin 1859). Darwin's genius laid the foundations of all modern bioscience and, arguably more than any other scientist, has had and continues to have the most significant impact. For example, Darwin's *theory of evolution* is proven over and over by scientists every day in routine laboratory experiments, as they clone genes and select for bacteria with resistance to specific antibiotics. Medically, every person who suffers from a bacterial infection and is treated with antibiotics is told by his or her doctor that it is crucial to continue the course of the antibiotics for the full ten days (or whatever time period is deemed sufficient)—even if the symptoms clear up far more rapidly. Why? Darwin's *theory of evolution* maintains that if there are small numbers of bacteria that are less sensitive to the antibiotic, and if they survive the initial treatment, then once antibiotic treatment is halted this more stubborn, resistant population may expand and grow and ultimately will be much harder to treat.

Darwin's *theory of evolution* is also painfully clear to many cancer patients; unfortunately cancer cells tend to undergo mutations, and once a random mutation occurs that selectively gives the cancer cell an advantage—such as becoming less sensitive or even resistant to a chemotherapeutic agent—these more dangerous cancer cells will then continue to grow and thrive until all of the remaining cancer cells retain this selective advantage and the cancer is no longer susceptible to treatment. In this way, while the chemotherapy may eradicate 99.99% of the tumor cells, the 0.01% that remain are incredibly hard to destroy. These are serious considerations for oncologists when they treat cancer patients and worry that they may develop resistance and suffer from a relapse. Undoubtedly the emergence of more transmissible COVID-19 variants in the midst of the 2020 pandemic also highlights Darwinism at work in a manner that the general public is forced to grasp.

Darwin's ideas are also firmly embedded in everyday life. Not only is human and animal behavior rife with the intense honing of evolution to produce the extraordinary sexual drive that serves to spread DNA and genes to future generations, but Darwin's principles are also inherent in the cross-bred food crops that we eat, the weeds we cull from our gardens, and the cockroaches and other pests of which we seek to rid ourselves. In short, it would not be an exaggeration to say that Darwin was about a hundred years ahead of his time—for it was not until the 1950s that scientists began to develop any solid molecular mechanisms to help explain Darwin's brilliant observations.

It is necessary to emphasize the importance of understanding the history of scientific discoveries, and how they have come about. Nobel Prize laureate and immunologist Peter Medawar wrote in a chapter about fellow Nobel laureate Jim Watson (in a chapter titled "Lucky Jim") in his book *The Strange Case of the Spotted Mice and Other Classic Essays on*

Science that "the history of science bores most scientists stiff," and that many creative scientists think that "an interest in the history of science is a sign of failing or unawakened powers." Medawar accurately realized the value of understanding the process of scientific discovery for scientists. However, it is clear that the history of science—with all of its convoluted processes of discoveries, which often occur with one step forward and another two steps going backward before ultimately marching on—must also be understood by students and the general public. Only once society fully understands the complexities and inherent messiness of science, and once society acknowledges the fogginess of initial discoveries until the dust clears—it is only then or often years later that the general public can really comprehend the limitations and science and the way that it works. Magic bullets are rare; when data comes in, it can sometimes be conflicting until things are resolved. The 2020 COVID-19 pandemic and evolution of scientific advice on dealing with it serve as a classic illustration and show how difficult it is to rush science. Understanding how science works leads to realistic expectations, but also highlights the significance of long-term investments in basic research.

As noted earlier, a recurrent theme in science is the idea that most scientific "discoveries" occur within the context of a solid foundation of previous findings, sometimes with leaps and bounds, but almost inevitably based on the scientific "bricks" laid by earlier pioneers. At times it is often the technologies—which may be technological rather than necessarily biomedical in nature—that are essential for the biomedical breakthroughs. Without X-ray crystallography, Franklin, Watson, and Crick would have been unable to resolve the structure of DNA. Without a technical means for measuring glucose serum levels, Banting and Best would have been unable to carry out their research that led to the discovery of insulin. But in addition to the biomedical advances made upon the availability of new technological advances, scientific ideas and discoveries are often not only built on a foundation of previous data, but many times "rediscovered." Indeed, not to detract from Fleming's hugely significant observation that mold on a Petri dish was capable of preventing bacterial growth—leading of course to the discovery of penicillin—it must nonetheless be noted that salves derived from moldy bread were reportedly used to wrap up wounds and skin infections by the ancient Egyptians, and moldy soy beans were used by the Chinese, whereas the Jews in Talmudic times used moldy corn for therapeutic purposes (Wainwright 1989). Thus, as scientists, it is necessary to always remember that there is a reason our scientific studies are known as *research.*

1 The Birth of Genetics*

In 1854, a monk with remarkable intellectual curiosity by the name of Gregor Mendel began to grow peas in the glasshouse of the St. Thomas Monastery in Austria. By 1865, Mendel had made a series of observations that ultimately changed the fundamental understanding of how traits are inherited by living organisms. He presented his key findings about what he termed "certain laws of inheritance" in 1865, and subsequently his studies were largely ignored by the scientific community for the next 35 years. Why? Consistent with the idea that the *timing* of scientific discoveries needs to be appropriate for those findings to be properly appreciated, Mendel's *laws of inheritance* preceded the large-scale acceptance of Darwin's *theory of evolution.* Indeed, it was not until the early 1900s that widespread knowledge of Mendel's *laws of inheritance* was propagated, and it took even longer until his ideas were fully accepted. At the turn of the twentieth century there were three new papers published, each of which *rediscovered* Mendel's *laws of inheritance.* As noted by Robin Marantz Henig in the prologue of her book *The Monk in the Garden*, "The explanation usually given for this curious turn of events is that the world wasn't ready for Mendel's laws in 1865, and that by 1900, it was" (Henig 2000).

Whether or not Mendel was a genius or simply a seasoned plant breeder who happened to be in the right place at the right time—an issue debated by some scholars and historians—is for the most part, irrelevant. What *is* significant is that he was clearly among the first researchers to observe and publish findings that showed that traits are inherited, and he outlined predictable and strict mathematical rules that govern the passing of individual traits from parent to offspring as discrete particulate units that exist in pairs in all individuals.

Perhaps one of the main reasons why Mendel's work remained relatively obscure until the early 1900s was that the entity known as a *gene* remained so nebulous in the absence of a molecular understanding of what that entity entailed. The lack of a firm understanding of how genetic material is passed from generation to generation—or more accurately, what that genetic material is comprised of—made it extremely difficult for contemporary scientists to accept Mendelian genetics. Indeed, a great debate erupted following the publication of Cambridge researcher William Bateson's 1894 book *Materials for the Study of Variation: Treated with Especial Regard to Discontinuity in the Origin of Species*, in which Bateson outlined 886

DOI: 10.1201/9781003202974-1

examples of discontinuous variation in heredity (Bateson 1894). Bateson, who upon reading Mendel's studies on the genetics of peas years earlier, reportedly felt that he had been "scooped" by Mendel and became one of the biggest advocates of Mendelian genetics. The idea that genetic traits could skip generations as a result of being *recessive*—meaning that the trait is only passed on to offspring if it is inherited from both parents (Figure 1.1)—was somewhat revolutionary in that it seemingly opposed some of the new ideas that had been emerging from modern statistics. Around the time of Mendel in the late 1800s Francis Galton discovered the statistical concept of "regression to the mean"; simply put, if a sample point is extreme when observing random variables, then additional points observed in the future will more likely be closer to the mean and are less likely to be outliers (Galton 1886). Galton calculated that Darwin's evolution must occur by larger, discontinuous steps rather than by small incremental ones to prevent regression back to the mean (Gillham 2001). The scientists who favored the notion that evolution was a smooth and *continuous* occurrence were known as biometricians. Bateson and those who supported the Mendelian model were convinced that only *discontinuity* could explain inheritance of many traits, and thus a raging battle was fought in a series of letters and counter letters published in the journal *Nature.*

By the early 1900s, however, more evidence in support of Mendel's ideas was coming from a different direction. In particular, two scientists, Walter Sutton and Theodor Boveri, contributed greatly to this enterprise. Sutton did significant research under the tutelage of the famous Edmund B. Wilson

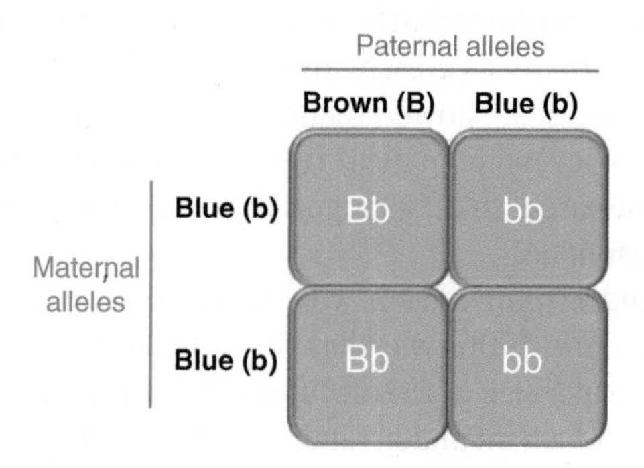

FIGURE 1.1 The Punnett square. The Punnett square, named after William Bateson's collaborator, Reginald Punnett, was devised by Punnett as an easy way to follow inherited traits. In the example here, in simplest form, each parent has two alleles, a capital **B** for brown eyes, and a small **b** for blue eyes. Since blue eyes are considered a recessive trait, the offspring must have two small **b** alleles to inherit blue eyes. In this case, the mother is blue-eyed with **bb** alleles, whereas the father is a brown-eyed person with one **B** allele and one **b** allele.

at Columbia University in New York, publishing "The Chromosomes in Heredity" (Sutton 1903) with the conclusion that chromosomes (which were now visible under the microscope by new cytological techniques) carry Mendel's hereditary material. The German cytologist, cell biologist, and zoologist Boveri had a remarkable career over which he made great discoveries, often relying on his zoological experience to make use of interesting systems to study. For example, he took advantage of fertilized sea urchin eggs and later the nematode *Ascaris megalocephala*, a parasite of the horse gut which later in his life infected him and may have caused his death (Maderspacher 2008). Boveri studied the centrosome or what he termed the "centrosoma" and documented its significance for cell division, as well as finding that the centrosome itself divides and organizes the surrounding cytoplasm in such a manner that the spindle fibers radiate from it and contact the chromosomes (Maderspacher 2008). Presciently, Boveri also published a lesser-known study in which he proposed that aberrant chromosomes might even be responsible for the generation of cancers in his 1914 book *Concerning the Origin of Malignant Tumors* (Boveri 2008). In any event, the work of these two scientists advanced the idea of Mendelian genetics greatly and led to what is known as the Boveri–Sutton chromosome theory.

Further support for Mendelian genetics came from the work of Nettie Stevens, who undoubtedly had to be an extraordinarily brilliant scientist to overcome the rampant misogyny of her era. Stevens was a geneticist who trained in the laboratory of Thomas Hunt Morgan, who was a famous fly geneticist and Nobel laureate for his own contributions to chromosomes and genetics at Bryn Mawr where Edmund Wilson was also a faculty member. In the course of her doctoral studies, Stevens received a fellowship to travel to Germany and further train with Boveri before completing her PhD. Upon returning to the US, she worked on mealworms (*Tenebrio molitor*) and observed that somatic cells of female mealworms contained 20 large chromosomes, whereas those of the male mealworm had 19 large chromosomes and 1 small one. She also found that exactly half the spermatozoa cells from the males contained nine large chromosomes and one small chromosome, whereas the other half had 10 large chromosomes. Her conclusion was that the eggs fertilized by the sperm with the 10 large chromosomes gave rise to female mealworms, and therefore the small chromosome dictated the generation of male mealworms. Her discoveries were further validated by Wilson when he looked at chromosome numbers in the insect *Hemiptera* (Wilson 1907), thus supporting the Mendelian genetics and providing a new twist to the mechanisms of heredity.

With support from the findings of the great cytologists and cell biologists, Mendelian genetics won the day. However, Mendelian genetics were ultimately combined with Galton's mathematical advances to yield new statistical methods and give birth to the modern field of genetics. All that was missing was an understanding of what comprised the particulate hereditary element within cells and chromosomes that allowed for the passage of traits from one generation to the next. Thus, the field of genetics was born through observations and careful experimentation, but without understanding the role of DNA in this process, scientists were still looking at the tip of the iceberg.

2 The Dawn of DNA*

In retrospect, it is easy today to connect the dots between Darwin's brilliant *theory of evolution*, Mendel's compelling observations on the mode by which offspring inherit traits from their parents, and the identification of deoxyribonucleic acid (DNA) as the genetic material that contains the blueprint of life. Back before the twentieth century, scientists knew that DNA existed, but its importance remained largely unrecognized. In 1869, a young Swiss doctor by the name of Friedrich Miescher decided to do research in Germany under the tutelage of Felix Hoppe-Seyler at Tubingen University. His interest was primarily in understanding the fundamental chemical composition of cells, and he intended to elucidate the principles of life within those cells (Dahm 2005). Miescher chose the white blood cells known as lymphocytes as his source of cells, because they were relatively easy to obtain. However, being unable to obtain sufficient quantities from lymph nodes, he switched specifically to the blood cells that migrated to wounds, using those isolated from the pus on fresh surgical bandages of patients.

At the time, the study of proteins was an exciting new area of research, and many researchers (wrongly) suspected that proteins were the genetic material that was passed from generation to generation, conferring traits on the recipient offspring. Accordingly, Miescher began his studies focused on proteins, but soon found that his background and the lab he worked in were not well-equipped to tackle the many issues that surfaced regarding the great diversity of proteins. It is estimated that there are ~35,000 distinct proteins expressed in most human cells, and this led Miescher to have serious concerns as to whether he would be able to properly classify them. However, as is frequently the case when well-trained and observant researchers keep their eyes and minds open when examining experimental results, Miescher found a material from the cells that tended to precipitate from the solution when an acid was introduced. The same material then dissolved again when a base was added to the solution (Miescher 1871). This initial observation was the first demonstration of the existence of DNA, the fundamental hereditary building block that eluded both Darwin and Mendel.

For all his scientific prowess, Miescher was not prolific in publishing his research (Dahm 2005), and the vast majority of his findings that were eventually brought to light have been revealed through his correspondence with his uncle Wilhelm His, who published Miescher's scientific

DOI: 10.1201/9781003202974-2

correspondence several years after the latter's premature death (His 1897). In these letters to his uncle, Miescher described the novel methods that he so painstakingly designed to first isolate the leukocytes from the surgical bandages of patients, ultimately relying on filtering the cells through a sheet to remove impurities and the cotton from the bandages, and taking advantage of gravity for a natural sedimentation of the cells. However, to separate and enrich the nuclei, the region of the cell that Miescher was convinced contained this unusual non-proteinaceous material (ultimately, the DNA), he had to engineer novel protocols that entailed, among other methods, long hydrochloric acid treatments at what he called *wintry temperatures* to avoid degradation of the molecules—fortunately northern Europe did not lack cold temperatures. Nonetheless, such protocols often took weeks and even months to complete. Ultimately, by using ether to extract lipids from his nuclei, Miescher found that further adding sodium carbonate to the solution containing the nuclei led to the isolation of what he termed *nuclein*, a substance that he maintained was not a protein due to its high level of phosphorus.

Miescher was clearly sufficiently savvy and astute to understand the significance of his novel discovery. He speculated that an increase in the level of nuclear materials (such as nuclein) might precede cell division in proliferating tissues (Dahm 2005). Indeed, he was so convinced of the physiological significance of nuclein that he promoted the idea that the nucleus should be defined by the presence of his newly identified nuclein, rather than merely on the grounds of morphology (as was typical at the time), which was a rather revolutionary concept (Miescher 1870).

The discovery of DNA, or nuclein as it was originally called by Miescher, is another superb example of serendipity in basic science. A researcher sets out to study one process—but observes something new and unexpected. It is human nature to plan, and even adhere to plans. Researchers are often unwilling to let go of fixed ideas, and even today, are frequently applauded for their ability to fixate on a given goal with dogged determination. But as demonstrated in this example by Miescher, and time and time again throughout the modern history of biomedical science, those researchers who are able to plan well, but also retain the flexibility to adapt and pursue promising new findings, have on occasion been known to make groundbreaking new discoveries.

In retrospect, it may seem odd to modern biomedical researchers that Darwin, Mendel, and indeed Miescher did not manage to connect the dots and collectively put the puzzle pieces together. Evolution, heredity and genes, and DNA, the molecule needed to tie it all together—all the information was present, so why didn't a consensus for the role of DNA emerge? Throughout the history of science, despite brilliant insights and unyielding persistence, it is not uncommon to experience circumstances where the element of timing may have been "off." Each of these individual discoveries was ahead of its time. Perhaps had communication technologies been more advanced and had the knowledge of these discoveries been more rapidly disseminated within the scientific community, the scientists might have collectively reached conclusions that essentially took an additional 60–70 years to come to fruition. However, in the absence of more advanced scientific techniques to study DNA, coupled with the slow scientific communication of the nineteenth century, DNA was destined to stay put in the nucleus and largely out of the limelight until its "rediscovery" in the 1940s and 1950s.

3 DNA as the Transforming Principle*

In the 1920s, a British researcher working for the health ministry in the United Kingdom named Fred Griffith coined a new term: *transformation*. Griffith was studying Pneumococcal bacteria, as he was interested in potentially generating a Pneumococcal vaccine, and he focused on two types of the bacteria. The first type had a smooth texture and it was protected from the immune system by its smooth, surrounding capsule. This smooth type of Pneumococcus was also lethal when injected into mice. The second type of Pneumococcal bacteria he worked on was rough and non-lethal when injected into mice, reportedly due to its induction of a significant immune response in the host animals. As expected, when Griffith heated and killed the lethal smooth bacteria before injecting them into the mice, the animals survived, similar to when they were injected with the live "rough bacteria." However, Griffith found it remarkable that when he injected the killed smooth strain into mice *together* with live bacteria of the non-lethal rough strain, the mice died (Figure 3.1). Why did the mice die? His conclusion was that a *transforming principle* from the dead smooth bacteria had *transformed* the live rough bacteria and rendered them lethal (Griffith 1928). These findings, more than anything else, had opened a new window into potential methods to study the composition and workings of genetic material, in an age when most researchers still firmly believed that proteins contained this information. Interestingly, Griffith, whose findings spurred on the discovery that DNA is the genetic material, was a researcher who was primarily interested in public health and infections caused by Pneumococcal bacteria. Indeed, at the beginning of his paper, he wrote the following: "The main point of interest, since the beginning of the inquiry, is the progressive diminution in the number of cases of pneumonia attributable to Type II pneumococcus." So while the focus of this book is that many great discoveries of medical value are made by basic scientists, often serendipitously, it is important to point out that when researchers are well-trained, qualified, and open-minded, sometimes the reverse can occur: in this case, Griffith set out with a very medical-related goal, but his findings became the spring-board for perhaps the most significant biomedical breakthrough of the twentieth century—that DNA is the hereditary material.

DOI: 10.1201/9781003202974-3

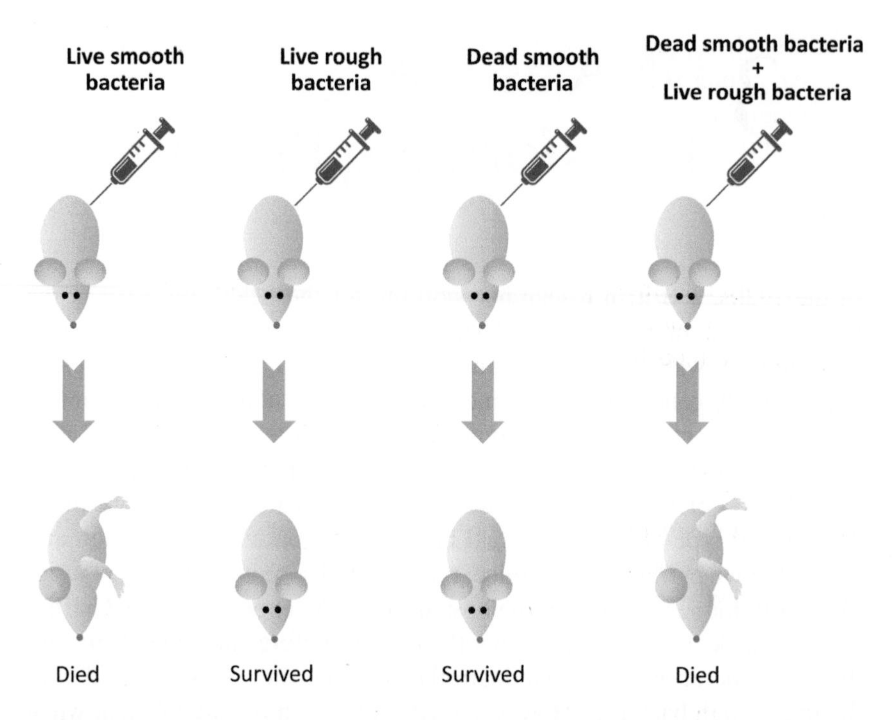

FIGURE 3.1 The transformative principle. The injection of mice with live, smooth (virulent) bacteria caused them to die, whereas the injection of live, rough, non-virulent bacteria did not kill the mice. When the smooth, virulent bacteria were first heat inactivated (killed) prior to being injected into the mice, the mice remained alive. However, when the heat-inactivated (killed) smooth, virulent bacteria were injected together with the live, rough, non-virulent bacteria, the mice died. Griffith inferred that something from the killed, smooth, virulent bacteria was capable of "transforming" the live, rough non-virulent bacteria to make them lethal, and he called this the *transformative principle*.

Oswald Avery should probably be credited with one of the key contributions to *transforming* (pun intended) how scientists viewed heredity and understood genetic material. Indeed, his 1944 landmark paper in *The Journal of Experimental Medicine* changed the trajectory of biomedical science (Avery, Macleod, and McCarty 1944). Avery, together with Colin MacLeod in his lab, began studying this novel concept of transformation, and their work showed that the material capable of transforming the bacteria was a white-colored precipitate that was sensitive to enzymes that digested DNA, or as it was called at the time, desoxyribose nucleic acid. Indeed, there were a number of indicators to suggest that DNA might be the transforming principle: (1) applying certain enzymes or select temperatures or ultraviolet light, all of which interfere with the integrity of DNA, similarly disrupted the unknown transforming principle, (2) the material that comprised the transforming principle generally failed to induce an immune response when injected into animals, whereas it would be expected that if the transforming principle were a protein its introduction into animals would lead to an immune response and/or antibody generation, (3) DNA and the transforming principle material behaved similarly when subjected to various biochemical treatments, such as centrifugation and separation in an electric field by electrophoresis. So while in retrospect it may be easy for scientists today to say: "If it walked like a duck (or like DNA), and talked like a duck, then it must be a duck," in reality, Avery was highly cautious with the conclusions he drew, and also acutely aware of the pushback he would likely receive from many of his colleagues in the scientific community. His conclusion in the paper was therefore rather tempered: "If the results of the present study on the chemical nature of the transforming principle are confirmed, then nucleic acids must be regarded as possessing biological specificity the chemical basis of which is as yet undetermined" (Avery, Macleod, and McCarty 1944).

In an interesting article by Matthew Cobb, the author maintains that despite Avery's overly cautious published statements, he was well aware of the significance of his findings and was more open about the long-term implications when discussing his results orally with other scientists (Cobb 2014). It is entirely possible that Avery's cautious interpretation of his findings later set the tone for the high accolades that Watson and Crick received for their DNA structure—whose major implication allowed for the reasonable speculation that DNA was uniquely structured to replicate itself. But to be fair to Avery, who unlike Watson and Crick did not receive a Nobel Prize (nor did he really receive full credit for his discovery—after all, even today Avery is not recognized as a household name in the biomedical world), his studies indicating that DNA was the transforming principle necessarily meant that DNA would need to be replicated so it could be passed on to every dividing cell.

Griffith and Avery may not have received their due credit for bringing forth evidence and promoting the notion that DNA serves as the genetic material that is passed on from generation to generation. However, the discovery of the transforming principle arose from both medically oriented vaccine research, and basic research. Clearly, a key lesson is that well-trained and talented biomedical researchers succeed when they maintain an open mind, no matter whether their initial goals were basic, translational, or clinically oriented.

similarly disrupted the unknown transforming principle, (2) the material that exhibited the transforming principle generally failed to induce an immune response when injected into animals, whereas it would be expected that if the transforming principle were a protein its introduction into animals would lead to an immune response and/or antibody generation, (3) DNA and the transforming principle material behaved similarly when subjected to various biochemical treatments, such as centrifugation and separation in an electric field by electrophoresis. So while in retrospect it may be easy for scientists today to say, "If it walked like a duck (or like DNA), and talked like a duck, then it must be a duck," in reality, Avery was highly cautious with the conclusions he drew, and also acutely aware of the pushback he would likely receive from many of his colleagues in the scientific community. His conclusion in the paper was therefore rather tempered: "If the results of the present study on the chemical nature of the transforming principle are confirmed, then nucleic acids must be regarded as possessing biological specificity the chemical basis of which is as yet undetermined" (Avery, Macleod, and McCarty 1944).

In an interesting article by Matthew Cobb, the author maintains that despite Avery's overly cautious published statements, he was well aware of the significance of his findings and was more open about the long-term implications when discussing his results orally with other scientists (Cobb 2014). It is entirely possible that Avery's cautious interpretation of his findings later set the tone for the high accolades that Watson and Crick received for their DNA structure—whose major implications allowed for the reasonable speculation that DNA was uniquely structured to replicate itself. But to be fair to Avery, who unlike Watson and Crick did not receive a Nobel Prize (nor did he really receive full credit for his discovery—after all, even today Avery is not recognized as a household name in the biomedical world), his studies indicating that DNA was the transforming principle necessarily meant that DNA would need to be replicated so it could be passed on to every dividing cell.

Griffith and Avery may not have received their due credit for bringing forth evidence and promoting the notion that DNA serves as the genetic material that is passed on from generation to generation. However, the discovery of the transforming principle arose from both medically oriented vaccine research and basic research. Clearly, a key lesson is that, as with [illegible] an open mind, no matter whether their initial goals were basic, translational, or clinically oriented.

4 The Structure of DNA Lends Itself to a Model for Its Duplication**

In 1953, shortly after the structure of DNA was solved by James Watson and Francis Crick, and based on information provided by Maurice Wilkins from X-ray crystallography performed by Rosalind Franklin, Watson and Crick wrote the following in a paper published in the journal *Nature*:

> The importance of DNA within living cells is undisputed. It is found in all dividing cells, largely if not entirely in the nucleus, where it is an essential constituent of the chromosomes. Many lines of evidence indicate that it is the carrier of a part of (if not all) the genetic specificity of the chromosomes and thus of the gene itself.
>
> **(Watson and Crick 1953)**

They later went on to note that since they propose that DNA serves as genetic material, one of its key functions must be to undergo some mechanism of self-replication, and they suggested the following (and today well-known) mechanism (Figure 4.1):

> Now our model for deoxyribonucleic acid is, in effect, a pair of templates, each of which is complementary to the other. We imagine that prior to duplication the hydrogen bonds are broken, and the two chains unwind and separate. Each chain then acts as a template for the formation on to itself of a new companion chain, so that eventually we shall have two pairs of chains, where we only had one before. Moreover, the sequence of the pairs of bases will have been duplicated exactly.

However unfairly Watson and Crick may have taken advantage of Rosalind Franklin's exceptional contributions to the discovery of the structure of DNA, and despite the idea that it may have been unlikely that Franklin felt any real animosity toward them in her sadly abbreviated lifetime (Maddox 2003), they clearly understood the immense implications of the findings. Indeed, in their landmark *Nature* paper, Watson and Crick essentially

DOI: 10.1201/9781003202974-4

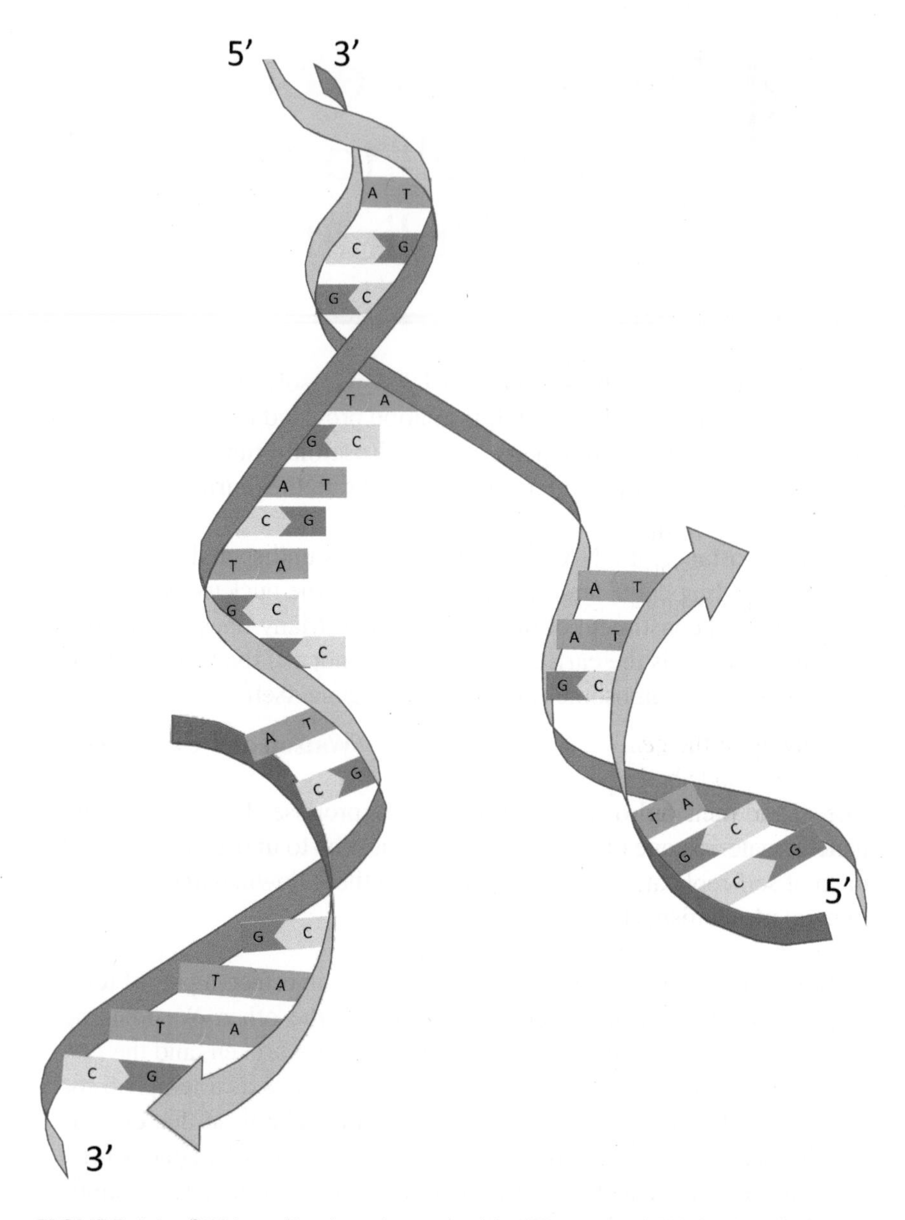

FIGURE 4.1 DNA replication. As posited by Watson and Crick, the DNA double helix is uniquely structured to facilitate its own replication. The two DNA strands are first separated from one another (bottom half of the diagram) by an enzyme called helicase. The separation of the DNA helix allows each of the two individual DNA strands to serve as a *template* for the synthesis of a new DNA counterpart strand by DNA polymerase, the enzyme responsible for filling in the appropriate nucleotides or *base pairs*. Note that each new DNA strand is synthesized in the opposite direction.

delivered two extremely important points: (1) they put forth strong evidence that DNA serves as the genetic material within chromosomes that is passed down from cell to cell upon division, and (2) they provided a very logical and highly accurate assessment of the mechanism by which the DNA double helix self-replicates so that a copy can be transferred to each daughter cell upon cell division. Indeed, much of what we ultimately know of the mechanism of DNA replication can be attributed to the work of another great twentieth-century biochemist and Nobel Prize laureate, Arthur Kornberg, whose contributions will be addressed in part in reference to the discovery of the polymerase chain reaction (PCR) in Chapter 6.

Today it is easy to take for granted these very fundamental observations on the role of DNA, but from the days of Mendel's genetic research on peas and Darwin's explosive revelation of the process of evolution, until the 1950s the molecular basis for genetics and evolution had remained largely obscure. Only in the mid-1950s when the complex structure of DNA was unveiled did scientists finally begin to develop a clearer picture of how genetics and evolution might occur at the molecular level. However, despite the enormous ramifications of discovering the structure of DNA, perhaps an even more earth-shattering scientific revelation—possibly the most brilliant and significant of all twentieth-century biomedical advances—was the solving of the *genetic code.* A decade later, the brilliant deciphering of the genetic code by Nirenberg, Khorana, Holley, and others may have ultimately had the single-most important impact on biomedical research and modern medicine, culminating with the birth of the modern era of molecular biology. Indeed, in retrospect, from the work of Darwin, Mendel, and Miescher, and on to Avery, Franklin, Watson, and Crick—to the cracking of the genetic code—the research and discoveries of these great scientists represents a continuum along a spectrum of biomedical advances, built layer upon layer, through the years and decades.

5 Tying It All Together
*tRNA, mRNA, Ribosomes, and the Genetic Code****

In retrospect it is easy to see the astonishing simplicity by which DNA serves as a cellular blueprint for all of the proteins in the cell. To some extent, it is almost surprising that it took scientists so long to identify DNA as the inherited genetic material that so elegantly and efficiently provides the remarkable adaptability necessary for the biogenesis of the tens of thousands of uniquely assembled and folded proteins that carry out the vast majority of functions needed for a cell to thrive. However, it is also possible to understand how researchers initially considered that DNA did not have sufficient variability to act as the hereditary material, with only four nucleotides comprising the DNA: adenosine (A), thymidine (T), cytosine (C), and guanine (G). Given this very limited number of nucleotides, it is not unreasonable to wonder how only four different types of DNA molecules might dictate the incredible variation found within living organisms. For these reasons, researchers were unwilling to give up on the notion that proteins, comprised of infinite permutations of the 20 distinct amino acids, were more likely candidates than DNA to comprise the transforming principle or genetic material that is passed down from generation to generation.

Beginning with the work of Miescher, followed later by the clever studies of Avery, and culminating with the determination of the structure of the DNA double helix by Watson and Crick, it gradually became clear and ultimately accepted that DNA is the hereditary material that makes up the units known as *genes*. However, the manner by which DNA could be converted from a set of blueprint-like instructions to actually passing along coded messages to the cell and inherited traits to daughter cells and organisms remained enigmatic and was the major constraint for biomedical researchers well into the 1960s, representing the next frontier of biomedical science.

The 1950s and 1960s were an exciting time in biomedical science, and the monumental discoveries made during this period culminated with our understanding of the *central dogma* of all biology. The central dogma maintains that DNA is transcribed into *messenger RNA (mRNA)*, which subsequently codes for the amino acids that make up specific proteins, and

DOI: 10.1201/9781003202974-5

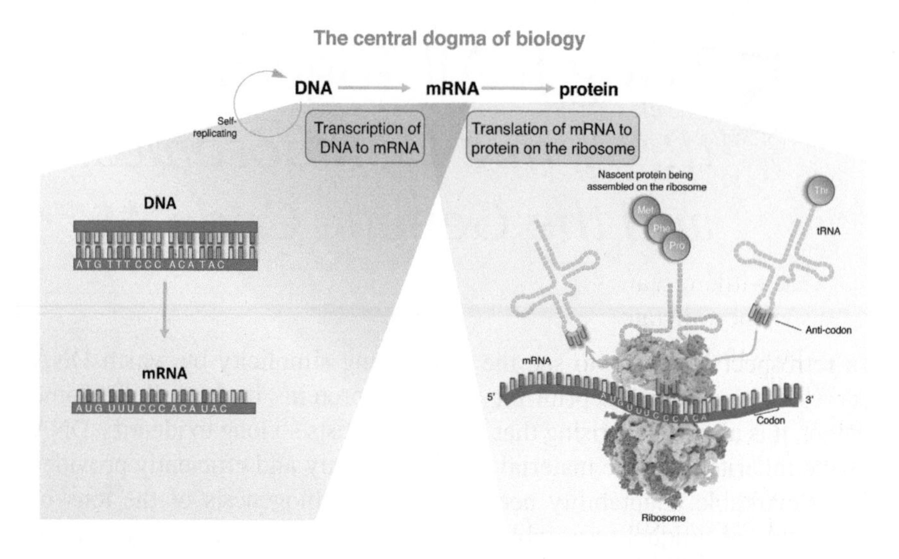

FIGURE 5.1 The central dogma of biology. In the 1950s, Francis Crick drew a similar schematic diagram, asserting (without yet fully understanding) that DNA is the hereditary material that self-replicates and is transcribed to mRNA before being translated to protein. Illustrated in more detail is the process of transcription of DNA to mRNA (left side) and a tRNA molecule carrying an amino acid into the ribosome to covalently attach it to the nascent polypeptide chain as specified by the mRNA (right side). While Crick recognized that this process was for the most part unidirectional (with the exception of retroviruses that can transform RNA to DNA), the involvement of transfer RNA (tRNA), ribosomes, and the actual genetic code signifying which amino acids are slated to be incorporated into proteins was not understood and would require extensive research by numerous laboratories and the better part of an additional decade before it was eventually clarified.

the deciphering of the code determines which of the 20 different amino acids are specified by each triplet of DNA nucleotides (Figure 5.1). These basic discoveries constitute the pillars of all of modern biomedical science.

Throughout the history of biomedical science, the most important advances have always come in "fits and spurts" contributed by numerous researchers around the globe, each exploring unique angles and perspectives, so that progress has not always occurred in a linear fashion. One may use a brick-on-brick analogy to describe how basic science advances: during an era in which the role and function of DNA were slowly characterized, bricks were laid around the entire foundation, almost haphazardly. However, within a decade or so, with the continued addition of bricks to the foundation, a very fine and solid structure began to arise. Indeed, the dramatic scientific advances came in parallel from so many different areas of research and so many different laboratories that assigning actual credit for many of the discoveries has been and still remains a daunting task.

Once it became clear that the DNA indeed contains the hereditary material of the cell, and that its structure lends itself to replication prior to cell division, biomedical researchers were left desperately trying to understand how the DNA is ultimately translated into the roughly 35,000 different proteins that the average human cell is estimated to have. A key problem was that while researchers envisioned that the order of the DNA nucleotides somehow dictated the specific sequence of amino acids incorporated into a protein, the DNA remained in the nucleus, whereas the proteins were reportedly synthesized outside of the nucleus on cytoplasmic organelles that were discovered by George Palade (Palade 1955). These cytoplasmic organelles where proteins undergo synthesis were functionally characterized by Paul Zamecnik (Littlefield et al. 1955) and termed *ribosomes* by Richard Roberts (Roberts 1958).

At this point in time, researchers in biomedical science were grappling with identifying all of the key players involved in the process of decoding the DNA and its genes into specific proteins. Today's high school students and university undergraduates are generally familiar with the famous central dogma of biology, which as noted earlier proposes a mechanism by which DNA self-replicates and is transcribed into an mRNA sequence (Figure 5.1). This is subsequently followed by the decoding of the mRNA into a string of attached amino acids that collectively comprise a specific protein, once it properly folds (Figure 5.1). However, both this proposed dogma and the identification of the key players involved in this DNA-to-RNA-to-protein cascade were still under intense investigation in the 1950s and 1960s. These studies eventually culminated with the brilliant work of Nirenberg and colleagues in the early 1960s, with the identification of the precise amino acids encoded by each of the 64 different combinations of DNA (or more directly, mRNA) nucleotides (Matthaei et al. 1962, Nirenberg and Matthaei 1961, Matthaei and Nirenberg 1961b, Matthaei and Nirenberg 1961a, Nirenberg et al. 1963, Martin et al. 1962).

Marshall Nirenberg began his research career as a young scientist at the National Institutes of Health in Bethesda, Maryland, and he was determined to work on an important biological question. The specific problem that he chose was that of protein translation and understanding the genetic code—studies for which he would later receive the Nobel Prize. Nirenberg was a very dedicated and ambitious researcher, but admittedly he did not have a strong background in molecular genetics, which made him a "dark horse" in the race to elucidate how the DNA nucleotides specify protein translation. Indeed, according to Matthew Cobb, who wrote an essay addressing the question of "Who discovered mRNA," several statements in the discussion sections of Nirenberg's published papers suggest that he may possibly have confused mRNA with ribosomes and/or ribosomal RNA (Cobb 2015). At the time, key scientific researchers in the field, including Watson and Crick, belonged to an exclusive "RNA Tie Club" that had 20 members in its ranks, with a single scientist-member representing each of the 20 amino acids. It is telling that Nirenberg was not invited and did not belong to this elite and selective club.

One of the key experiments that Nirenberg directed and performed that changed the way he was perceived by others in the field was one that he carried out together with a postdoctoral fellow who worked with him, Heinrich Matthaei (Matthaei et al. 1962). Nirenberg, along with many other researchers at the time, still did not really fully understand the relationship between mRNA and proteins, and he aimed to know whether the sequence of RNA could serve as a genetic template and dictate the sequence of amino acids that string together to comprise a specific protein. Fortunately, a major technical advance in biochemistry made by the groups of Severo Ochoa (Beljanski and Ochoa 1958) and Paul Zamecnik (Littlefield et al. 1955) was extremely helpful to Nirenberg. These groups were able to set up unique methods for the translation of proteins in a *cell-free system* using the cytoplasm extracted from bacteria as a source for all the ingredients needed to complete the process, and Nirenberg took full advantage of these new systems.

Nirenberg and Mathei proceeded to use this type of cell-free *in vitro* system and tested the addition of select RNA polynucleotides (strings of RNA nucleotides) to determine whether the sequence of this RNA could dictate the composition of a nascent polypeptide (protein) chain that was generated in the cell-free assay. Their experimental system was relatively simple: using the bacterial cytoplasm as their source of the ribosomes (the organelle where proteins undergo synthesis), tRNA (for delivery of the amino acids as specified by the RNA codons), and all the other "goodies" needed for protein translation, the two researchers added their cytoplasmic extract to 20 different test tubes, 1 for each of the 20 different amino acids. In each test tube, however, only 1 of the 20 amino acids was radioactively labeled, although the other 19 non-labeled amino acids were also provided. When Nirenberg and Matthaei then added their synthetic RNA to each of the test tubes, they started with poly-uracil (essentially a string of uracils, known as poly-U), and they tested the incorporation of the "hot" or labeled amino acid from each of the 20 test tubes into a growing protein (Figure 5.2). Only the vial containing the radioactive amino acid phenylalanine displayed incorporation into a protein when synthetic poly-U was used as the RNA template (or mRNA). Thus Nirenberg and Mathei had

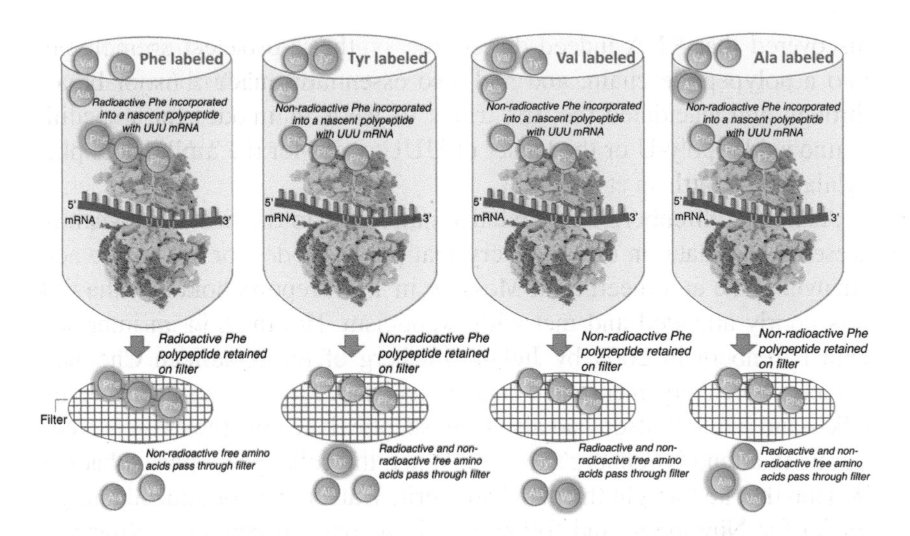

FIGURE 5.2 Elucidation of the genetic code by Nirenberg and Mathei. This depiction of the classic experiment by Nirenberg and Mathei shows how they were able to systematically assess all 64 nucleotide triplet combinations (codons) to derive the amino acid for which each triplet codes. In this abbreviated explanation, the researchers derive the code for the triplet UUU, using a string of UUU codons as their mRNA to be tested. They supplied ribosomes, amino acids, tRNA, and all of the ingredients and machinery required for protein translation in a test tube (*in vitro*), but for each triplet combination that they used as mRNA (in this example just the UUU) they had 20 different reactions, 1 for each different amino acid. Each reaction had 19 unlabeled amino acids, and only 1 radioactive amino acid. Since UUU codes for the amino acid phenylalanine (Phe), in the reaction on the far left *only* the radiolabeled Phe will be incorporated into a polypeptide chain bound to the ribosomes. Accordingly, when the components of the reactions are passed over a filter paper, only the small components will pass through and the larger ones will be retained on the filter. In this case, all of the other amino acids (that are non-labeled in this case) except Phe passed through the filter, whereas the radiolabeled Phe, which binds to the mRNA and is thus attached to the large ribosomes on which the mRNA is positioned, was captured by the filter, as the ribosomes are too large to pass through. For this reason, the filter paper will retain high levels of radioactivity, thus indicating that UUU codes for Phe. In the subsequent reactions portrayed, the UUU still caused chains of Phe to be synthesized that stick to the filter paper through their attachment with the ribosomes, but since in each of these reactions a different amino acid is labeled (and not Phe), no radioactivity adhered to the filter. Such a system was repeated for each of the 64 triplet combinations, each time using 20 reactions with a different amino acid radioactively labeled.

discovered that RNA indeed directs the synthesis of select amino acids into a polypeptide chain, and had also essentially made a major breakthrough and elucidated the first triplet of RNA that codes for a specific amino acid—poly-U or the triplet of UUU codes for the amino acid phenylalanine (Matthaei et al. 1962).

As noted, Nirenberg was a relative newcomer to the field, and when he presented his data on the discovery that poly-U codes for the amino acid phenylalanine at a meeting in Moscow in 1961, reports hold that the talk was poorly attended and met with skepticism. However, in an interview with Nirenberg in 2009 by Judah Ginsberg of the American Chemical Society, Nirenberg mentioned that he had met James Watson before his talk—and while Watson had also been skeptical, he nonetheless sent a colleague to attend Nirenberg's lecture. When the colleague reported back to Watson that he thought the data had merit, Watson then organized a larger forum for Nirenberg, and following his second presentation, Nirenberg said "The reaction was incredible. It was a standing ovation … but for the next 5 years I became like a scientific rock star."

Despite the major breakthroughs, including the concept that RNA is the intermediary messenger for protein translation, and the notion that specific RNA nucleotide sequences dictate amino acid incorporation into proteins, this was only the tip of the iceberg. Many key questions remained unresolved, and the rest of the code remained elusive. For example, while a string of poly-U RNA nucleotides clearly coded for a polypeptide comprised of a string of phenylalanines, it was still not known how many uracil nucleotides coded for a single phenylalanine. Even once it had been discovered that a *codon* that codes for a single amino acid is comprised of just three nucleotides, it was not known whether the triplet nucleotide codons were consecutive, with the second three nucleotides (or codon) coming precisely after the first three in the mRNA sequence, or whether there needed to be a gap of one or more nucleotides before the codon for the next amino acid began (Figure 5.3). Slowly and meticulously, these questions were resolved in favor of a consecutive, continuous, unpunctuated code of three nucleotides with no gaps in between codons (Figure 5.3).

Nirenberg raced ahead taking advantage of new technologies, some of which were developed by fellow 1968 Nobel Prize winner Har Gobind Khorana who also contributed to the deciphering of the genetic code, including the ability to very precisely synthesize RNA triplets representing all of the 64 possible codons (4 nucleotides × 4 nucleotides × 4 nucleotides = 64 combinations). Using the system depicted in Figure 5.3 and all of the 64 possible mRNA triplet combinations, Nirenberg worked

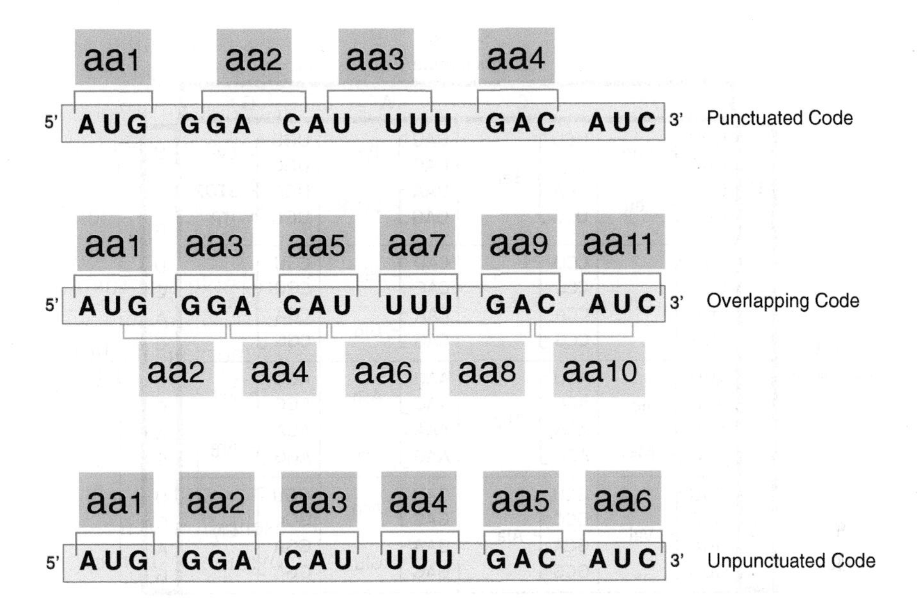

FIGURE 5.3 The genetic code is an unpunctuated code. Three potential ways in which mRNA codons could be read at the ribosome: (1) as a *punctuated code* where there are gaps between the translated codons (top example), (2) as an *overlapping code* where nucleotides might belong to two different translated codons (middle example), and 3) as an *unpunctuated code* where there are no gaps or overlap between translated codons (bottom example). aa stands for amino acid.

with Phillip Leder to analyze the outcome for all possible combinations. They discovered that only 61 of the 64 possible combinations of mRNA codons actually code for an amino acid, with three triplets serving as *stop codons* that dictate the conclusion of protein translation and release of the protein from the ribosome, a process known as *termination* (Leder et al. 1963, Nirenberg and Leder 1964). What this also meant is that since there are only 20 amino acids but 61 different codons, there are multiple triplet combinations that can specify incorporation of the very same amino acid. Indeed, in some cases up to six different codons dictate coding of the same amino acid (Figure 5.4).

Despite the undeniable genius of Nirenberg in solving the puzzling concept of the genetic code and translation of specific mRNA sequences into polypeptides, at the time Nirenberg and his co-investigators did not completely understand the mechanism of translation—in particular, precisely *how* the various RNA players were involved. This in no way detracts from the Nobel-winning achievement of conclusively defining exactly which amino acids were specified by each of the 64 codon combinations, but

First nucleotide	Second nucleotide: U	C	A	G	Third nucleotide
U	UUU, UUC – Phe UUA, UUG – Leu	UCU, UCC, UCA, UCG – Ser	UAU, UAC – Tyr UAA, UAG – **STOP**	UGU, UGC – Cys UGA – **STOP** UGG – Trp	U C A G
C	CUU, CUC, CUA, CUG – Leu	CCU, CCC, CCA, CCG – Pro	CAU, CAC – His CAA, CAG – Gln	CGU, CGC, CGA, CGG – Arg	U C A G
A	AUU, AUC, AUA – Ile AUG – *Met	ACU, ACC, ACA, ACG – Thr	AAU, AAC – Asn AAA, AAG – Lys	AGU, AGC – Ser AGA, AGG – Arg	U C A G
G	GUU, GUC, GUA, GUG – Val	GCU, GCC, GCA, GCG – Ala	GAU, GAC – Asp GAA, GAG – Glu	GGU, GGC, GGA, GGG – Gly	U C A G

FIGURE 5.4 The genetic code is redundant. The table depicts the 64 different combinations of codons. Sixty-one codons effectively translate into 20 amino acids with considerable redundancy, whereas 3 codons effectively terminate protein translation and are known as *stop codons.*

nonetheless, it took a "village of scientists" to ultimately work out the complex mechanisms of the pathway.

There is some evidence that Nirenberg, who used synthetic RNA as the template or what we now know as mRNA, did not clearly distinguish between the different types of RNA required for the process of translation. These include the ribosomal RNA (rRNA) that has an important catalytic role in the translation process at the ribosome (rRNA essentially comprises a part of the ribosomal machine), the transfer RNA (tRNA) that serves as a scaffold to carry each of the 20 amino acids into the ribosome so that the amino acids can be added to the growing polypeptide chain, and the messenger RNA (mRNA) that specifies which amino acids are to be added to the nascent protein and in what order. Indeed, Matthew Cobb notes that "As late as December 1960, Nirenberg's diaries show that he was still toying with the idea that protein synthesis took place on the DNA molecule, something most of the scientific community had abandoned years earlier" (Cobb 2015).

However, one can hardly fault Nirenberg because during this exciting time, even the famed members of the prestigious and elite "RNA Tie Club" were still grasping at straws trying to work out the details of protein translation. After all, the great Francis Crick had been a proponent of the flawed hypothesis that held that all of the ribosomes in the cell were heterogeneous, maintaining that each ribosome served as a factory for the synthesis of a single type of protein from a single gene or RNA. He called this the "One gene, one ribosome, one protein hypothesis." Only once it had been proved that individual ribosomes contained very little heterogeneity, and that the rRNA from individual ribosomes showed little difference in its size, did Crick modify his model in 1959 to suggest that perhaps only part of the rRNA serves as the template to direct protein synthesis. It is important to credit Crick and other scientists who put forth incorrect, albeit instructive models, but in the wake of emerging evidence were willing to modify their ideas, precisely as good scientists should do. Indeed, just as important as the ability to propose scientific models is the willingness of researchers to display flexibility and the willingness to revise their models when new scientific evidence comes to light.

It is a fascinating anecdote that the actual discovery of mRNA took place almost in parallel with Nirenberg's initial identification of the first part of the genetic code. In other words, Nirenberg had begun to work out the language of protein translation without fully understanding the syntax of the language. Arguably, elucidation of mRNA as the real template for the genetic code was just as important as the code itself; the key to all advances in molecular biology well into the twenty-first century stems from the ability of researchers to manipulate complementary DNA (cDNA)—the DNA that corresponds to the mRNA template—to generate protein expression in mammalian cells, or protein synthesis in prokaryotic cells for the purpose of purification and *in vitro* analysis. All of these studies, which today are fairly routine in most biomedical laboratories, require a complete understanding of the basis of the central dogma of biology; simply knowing the key to coding would not be enough to support the dramatic scientific advances that are derived from our understanding of molecular biology.

In his essay "Who Discovered Messenger RNA?" Matthew Cobb notes that the very first researcher to correctly publish the sequence of events that essentially comprise the central dogma was Andre Boivin who wrote in 1947, albeit in a French-language journal, that "the macromolecular desoxyribonucleic acids govern the building of macro-molecular ribonucleic acids, and, in turn, these control the production of cytoplasmic enzymes" (Cobb 2015, Boivin 1947). Over time, a series of researchers that included Raymond Jeener (who first proposed that nuclear RNA might, in association with ribosomes in the cytoplasm, facilitate protein synthesis), the French team of Francois Jacob and Jacques Monod (who were later awarded Nobel Prizes for their work on understanding bacterial gene expression), and the trio of Sydney Brenner, Francis Crick, and

Jim Watson, along with many other researchers, all moved the ball forward, step by step, slowly but surely, until it eventually became clear that a short-lived RNA derived from the DNA in the nucleus is transported to ribosomes in the cytoplasm where the translation of proteins occurs. This brick-on-brick building of the central dogma highlights the necessity of assembling so many different researchers, working on so many diverse angles related to protein translation (and in the case of Jacob and Monod, even in more "peripheral" research that was not directly focused on protein translation and the genetic code), to tie everything together. It took a village to discover mRNA and ribosomes and elucidate the relative homogeneity of the latter (refuting Crick's "one gene, one ribosome, one protein" hypothesis), and ultimately to solve the mystery of the genetic code.

In the background of the work on mRNA, ribosomes, and the elucidation of the genetic code, there was still an important missing piece of the puzzle; how were the specific amino acids that were coded by the mRNA coupled to the growing polypeptide chain? Francis Crick had hypothesized an enigmatic "mystery molecule," and a large body of work led by another Nobel Prize winner, Robert Holley, identified, purified, and discovered the sequence and function of this new molecule, known as transfer RNA. Twenty distinct tRNA molecules have the remarkable function of each binding to a specific amino acid (each of the 20 individual amino acids) while at the other end of the molecule the tRNA recognizes the specific mRNA triplet or codon. In this way, the tRNA molecules serve as a molecular bridge to connect the correct amino acid to the site of the ribosome and mRNA so that it may be coupled to the growing protein chain. With scientific advances coming from parallel sources at a fast and furious pace (at least relative to scientific progress) from all different angles, not only was the genetic code elucidated, but also the remarkable mechanism by which proteins are synthesized according to the explicit mRNA instruction code was also unveiled.

So why are these discoveries so essential to our knowledge base today? Without a doubt, the central dogma and comprehension of the genetic code are hugely important; indeed, understanding the basis of protein translation may well be the single most significant discovery of the twentieth century in biomedical sciences. Arguably, the solution of the genetic code may be the greatest biomedical discovery *ever*. Why is this the case? It is because the basis for almost all of molecular medicine lies with these discoveries. Every genetic illness can be explained by mutations that affect DNA (genes), which in turn impact RNA transcription, thus altering the composition of the translated protein and its function within the cell, ultimately causing disease.

TABLE 5.1
Select Diseases Caused by Abnormal Proteins

Name of Disease	Protein Affected	Disease Manifestations
Hemochromatosis	Transferrin receptor 2 and other proteins	Excessive iron uptake
Hermansky–Pudlak syndrome	AP-3 proteins, HPS proteins	Oculocutaneous albinism, pigmentation problems, and effects on lysosome-related organelles
Huntington's disease	HTT protein	Neurodegenerative disorder, coordination issues, dementia
Joubert syndrome	Cep290, Arl13b, and a variety of proteins related to primary cilia	Developmental disorder affecting the central nervous system
Lynch syndrome	MSH2 and other proteins	Hereditary nonpolyposis colorectal cancer
Menkes disease	ATP7A protein	The ATP7A copper transporter protein is defective leading to copper deficiency, growth defects, and central nervous system deterioration
Morquio syndrome	Galactosamine-6 sulfatase	A lysosomal storage disorder and failure to degrade mucopolysaccharides, leading to impaired physical growth, heart disease, and a short lifespan
Niemann–Pick disease	NPC1, NPC2, and other proteins	Sphingomyelin accumulates and fails to undergo degradation in lysosomes, leading to varied severe neurological symptoms
Polycystic kidney disease	PKD1 and PKD2	Abnormal renal tubules and multiple kidney cysts
Warburg Micro syndrome	Rab3GAP protein	Developmental disabilities, microcephaly, microcornea, congenital cataracts

Understanding how individual mutations in genes lead to the generation of aberrant proteins and impact disease is certainly crucial for biomedical science. Indeed, the brilliant studies of Linus Pauling in the late 1940s first demonstrated that an abnormal protein can cause disease (see Chapter 9). The elucidation of the mechanisms of protein translation and the solution of the genetic code continue to have important ramifications for scientific research and new advances to this day. There is a long list of known genetic diseases caused by mutations, deletions, or other aberrancies in human DNA, leading to mutant or truncated proteins and ultimately, disease (see Table 5.1 for examples). Biomedical researchers continue to have a major focus on understanding the function of the ~35,000 proteins per mammalian cell (possibly as many as 100,000 proteins if one counts the various isoforms). For example, researchers strive to understand which of these proteins impacts specific pathways and diseases, and how they interact with one another to carry out the functions critical to the life of a cell. These essential questions are currently being addressed by scientists around the globe. The way researchers can properly study these proteins is with the knowledge gleaned on how to artificially create them *in vitro* (in a test tube) to study, or by introducing them into cells after engineering the proteins to have various amino acid substitutions, allowing scientists to determine how the altered proteins impact cellular function. In modern biomedical science, it has become a common procedure to "knock-down" the expression of specific proteins in the cell to assess how the cell fares in their absence. The rapidly accumulating knowledge of the function of tens of thousands of proteins since the cracking of the genetic code has revolutionized biomedical research, and the massive impact on improving human health cannot be underestimated. However, the ability to fully take advantage of these advances and leverage them toward curing disease still awaited additional major scientific discoveries, including new technologies for amplifying and manipulating DNA and genes.

6 The Practicality of PCR
*The Technology that Drove the Biotechnology and the Molecular Biology Revolution****

Throughout the history of science, there have been times when dramatic advances were made in understanding the molecular world, but a novel, ingenious, and practical way of thinking was required to fully capitalize on this new knowledge. In short, scientists are driven by a desire to better understand the world around them, but scientific pursuit of knowledge does not necessarily follow the most practical approaches. Sometimes scientific innovation needs a more businesslike or even commercial approach to transform understanding and render that knowledge useful. Such is the case with regard to the discovery of the *polymerase chain reaction*, also known as *PCR*.

As the elucidation of the genetic code was unfolding, research was also proceeding toward an understanding of the mechanisms by which DNA is replicated. Watson and Crick had indeed provided important clues as to how DNA double helices might unravel so that once separated from one another, each of the two DNA intertwined strands might serve as a template for the generation of a new, parallel strand of DNA. This of course allows for a dividing cell to provide both mother and daughter cells with the blueprint of life—their own set of DNA containing the hereditary material in the form of chromosomes and genes. In parallel, another Nobel Prize winner, Arthur Kornberg, discovered the protein known as DNA polymerase, an enzyme possessing the unique function of generating a copy of the very DNA whose replication was proposed in the Watson–Crick model (Lehman, Zimmerman et al. 1958, Bessman et al. 1958, Lehman, Bessman et al. 1958). Kornberg's brilliance was apparent, not just by virtue of his Nobel Prize-winning discoveries, but also from his prophetic understanding of how science advances, for he famously said: "No matter how counterintuitive it may seem, basic research has proven over and over to be

DOI: 10.1201/9781003202974-6

the lifeline of practical advances in medicine. Without advances, medicine regresses and reverts to witchcraft" (Kornberg 1959a).

In the 1950s when Kornberg began to address this problem of how the DNA makes a copy of itself prior to cell division, the mechanisms of DNA replication were not understood. It is important to emphasize that in addition to the immense significance of determining how human cells transfer their genetic material to the next generation, understanding DNA replication would turn out to be an absolutely crucial factor that led to the molecular biology revolution and advances in genetic engineering that arose in the 1970s. This revolution has allowed researchers to easily amplify almost any segment of DNA or gene, including human DNA, and to shuttle it into cells for the purpose of studying the proteins that it encodes. It has allowed radical new methods for the treatment of genetic diseases, and it has facilitated the generation of transgenic animals that express modified or truncated proteins, or even lack the expression of a specific protein altogether. Undoubtedly, the discovery of DNA polymerase and the ability of researchers to replicate DNA is crucial to modern medicine, enabling routine laboratory diagnostic tests, including testing for viral infections through the amplification of DNA. Copying, amplifying, and thus identifying viral DNA, including that derived originally from COVID-19 RNA, are diagnostic procedures that rely heavily on this knowledge. Indeed, practically every modern biomedical research laboratory today amplifies and manipulates DNA fragments to gain knowledge of the function that these genes encode. Accordingly, today's scientific research and many of the key medical diagnostic tests currently in use absolutely depend on the foundation of great biomedical discoveries from the middle of the twentieth century.

Despite the vast knowledge of DNA replication and the magnitude of this breakthrough in understanding the process, by itself it was nonetheless insufficient to revolutionize biomedical science. As noted above, a little something was still missing—a practical approach to capitalize, or perhaps even commercialize and industrialize the use of DNA amplification technologies. That approach turned out to be PCR. Biotechnology is often driven by a simple pair of words: "*So what*?!" *So what* if we know how DNA is replicated? *So what* if we understand how proteins are translated from mRNA on ribosomes? *So what*, because without some way to take advantage of these discoveries and actually derive some practical *use* from them, aside from their somewhat limited academic perspective these discoveries do not have a major impact on people's day-to-day lives. Into this partial void came the ingenuity of researchers whose thinking was couched in more practical terms.

Although the idea of being able to synthesize and amplify DNA in a test tube by utilizing the cell's own machinery can be traced back to Nobel Prize winner Har Gobind Khorana and his co-workers in a 1971 paper (Kleppe et al. 1971), much of the credit for the practical development of PCR has been awarded to Kary Mullis, who won the Nobel Prize years later in 1993. Building on Arthur Kornberg's identification and purification of DNA polymerase, the enzyme that copies a single strand of DNA once it has been unwound and separated from its paired strand, Kjell Kleppe, and Khorana and their co-workers wrote:

> The DNA duplex would be denatured to form single strands. This denaturation step would be carried out in the presence of a sufficiently large excess of the two appropriate primers. Upon cooling, one would hope to obtain two structures, each containing the full length of the template strand appropriately complexed with the primer. DNA polymerase will be added to complete the process of repair replication. Two molecules of the original duplex should result. The whole cycle could be repeated, there being added every time a fresh dose of the enzyme.

Essentially, Kleppe and co-workers had completely outlined the method by which DNA could be replicated *in vitro*, 22 years prior to Mullis' Nobel award. But as often is the case in science, the devil is in the details.

In his book *Making PCR: A Story of Biotechnology*, Paul Rabinow argues that although the concept of synthesizing DNA was around since Khorana's time, nonetheless from that time until Mullis developed the idea further at his company Cetus, not a single scientist applied for a patent (Rabinow 1996). Indeed, Rabinow cites a former Cetus scientist, Norman Arnheim, as saying: "Conception, application and development are all scientific issues—invention is a question for patent lawyers." Rabinow further notes that "Mullis conceived of a way to turn a biological process (polymerization) into a machine; nature served (bio)-mechanics."

Many researchers in the mid-1980s were reluctant to credit Mullis with the invention/discovery of PCR; indeed, in 1989 the journal *Science* named DNA polymerase as the "Molecule of the Year" and discussed PCR in great detail, but studiously avoided any mention of Mullis in this article (Guyer and Koshland 1989). Whether this represented a degree of animosity or at least disregard toward those involved primarily in the commercialization of scientific discoveries, or whether it reflected direct contempt for Mullis himself—who later became drawn toward anti-science conspiracy theories regarding HIV/AIDS and climate change—is not altogether certain. It nonetheless remains difficult to argue that Mullis did not play a key role in the revolution of modern biomedicine. One might submit that PCR would have been discovered or engineered in any case, with or without Mullis, but such arguments can be made for every great discovery—great discoveries are all about the *timing*.

So how does PCR actually work? The process, which Mullis first worked out in the early 1980s, was fairly straightforward (Figure 6.1): if one wants to replicate and amplify a particular stretch of DNA, then the following steps are necessary: (1) heat the double-helix DNA to *denature* it

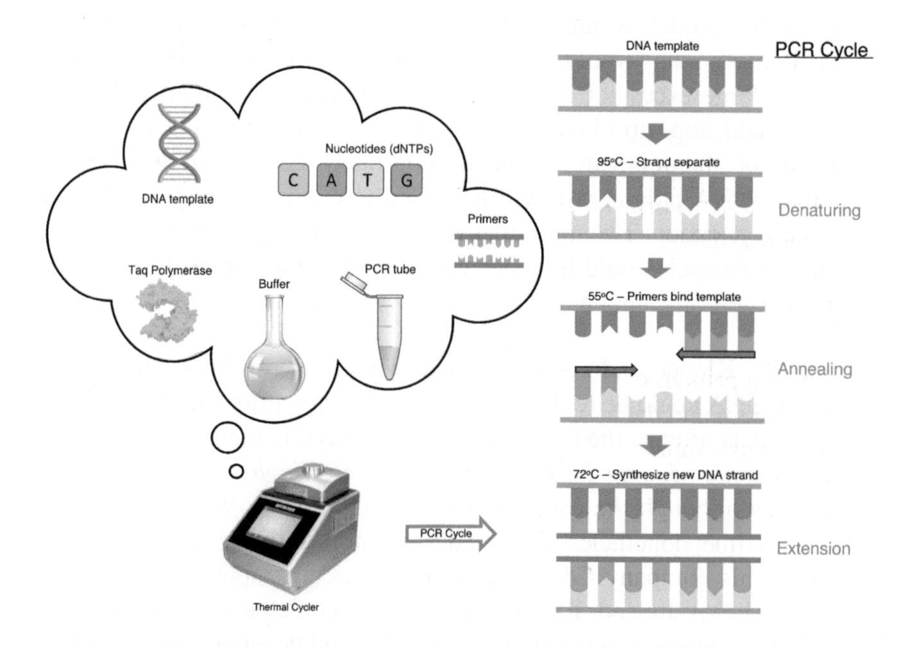

FIGURE 6.1 PCR amplification of DNA. A double-stranded DNA template (for example, a gene) to be amplified is added to a PCR test tube with the following "ingredients": excess nucleotides known as dNTPs that are needed to comprise the amplified DNA strands, short DNA oligonucleotides that match the ends of each of the DNA strands to be amplified (known as *primers*), Taq polymerase enzyme to synthesize the DNA by adding individual dNTP nucleotides to each strand, and reaction buffer in which the synthesis takes place. The ingredients in the PCR test tube are then inserted into a thermal cycler, which first heats the tube to 95°C to induce unwinding and separation of the double-strand DNA (called *denaturing*). The thermal cycler then cools down to about 55°C and thus allows the primers to again bind to each end of the template (called *annealing*), before heating back up to 72°C so that the Taq polymerase that functions at high temperatures (like those of hot springs) can fill in the missing dNTP nucleotides (a process called *extension*). This cycle is repeated about 20–40 times to generate a huge number of copies of the original template (or gene) that may then be used for genetic engineering and other purposes.

and separate the two attached and complementary (mirror-imaged) strands of DNA from each other, (2) cool down the DNA and add DNA polymerase and ample quantities of all four DNA nucleotides along with short mirror-image *primer* DNA pieces to bind to the template strand (the strand to be copied) and jump-start the process, allowing the reaction to proceed so that the DNA polymerase enzyme fills in the correct nucleotides, (3) after this cycle of synthesis, heat up the DNA to again separate all of the attached DNA strands (this time there will now be twice as many DNA double helices), (4) repeat the process to achieve exponential replication of the DNA. It is the short primer DNA piece that provides the specificity for the reaction and is the key to diagnostic tests. For example, if scientists want to determine whether a person has been infected with a virus, such as COVID-19, active infection would mean that there is viral DNA within the human cells (note: although COVID-19 is an *RNA virus*, its genetic RNA material is converted to DNA by the enzyme *reverse transcriptase*). Albeit, the DNA of viral origin is proportionally a very small amount out of the total cellular DNA. Accordingly, after taking a nasal swab of cells, and purifying DNA from that sample, scientists would need to greatly amplify the DNA whose origin was the virus (if it is present) to be able to detect it, thus determining whether the patient is infected or not. To this end, PCR is used with a primer, or possibly several different primers, to amplify DNA of viral origin and determine its presence in the sample.

Two things rapidly made this simple PCR process both feasible and effective: (1) the development of automated machines (known as *thermocyclers*) that heat up the DNA to 95°C for denaturing and separation of the strands, and (2) the identification and purification of a DNA polymerase that is derived from the bacteria known as *Thermus aquaticus* (fondly known as *Taq polymerase*), which is found growing in bacterial "mats" in the hot springs of Yellowstone National Park in Wyoming. Given its unique residency in bacteria that thrive in hot springs, it was perhaps not surprising that *Taq polymerase* turned out to be remarkably resistant to the very high temperatures needed for DNA denaturation. This heat-resistant and highly stable DNA polymerase greatly facilitates the process of PCR, because during the cycles of DNA denaturation at high temperatures, ordinary DNA polymerases would be damaged by the extreme temperatures, whereas the *Taq polymerase* from *Thermus aquaticus* remains perfectly functional. This further highlights how a basic science discovery—one that was made purely in the course of curiosity-driven research by Thomas Brock of the University of Wisconsin in Madison and had no perceived biomedical rationale at the time of its discovery (Brock 1997)—has become the backbone for genetic and viral PCR diagnostic tests, and of course it has revolutionized the entire field of forensic science, as well as biomedical research.

7 Genetic Engineering and Beyond

*From Animal Models to Silencing RNA***

How did the understanding of the genetic code and the rapidly accumulating knowledge of genetic engineering techniques alter the course of research? The most obvious way is that it provided a means for researchers to artificially induce the expression of any given protein in the cell. This technique involves the assembly and inclusion of a specific gene (or a segment of DNA) into an artificial *vector*, or *plasmid*. Plasmids, which are widely commercially available today, are circular pieces of DNA frequently used to "express" a given piece of DNA or gene in either mammalian or bacterial cells in the form of a translated protein. Plasmids typically have a DNA backbone that includes a *promoter* and other DNA segments that facilitate expression of the inserted DNA/gene that researchers wish to express (Figure 7.1). Plasmids are introduced into cells growing in culture in a process called *transfection*. In the process of transfection, there is a facilitated entry of the plasmid into the cell through the plasma membrane, and the DNA or gene that has been inserted or *subcloned* into the plasmid then undergoes transcription and translation leading to *over-expression* of the protein coded for by the newly introduced DNA. The reason it is considered over-expression is that the cell already (in most cases) has its own copy of that DNA and already expresses the protein that it codes for, but the transfection of the plasmid induces expression of much higher levels of that same protein. It is, however, possible to over-express the protein of interest at high levels even in cells that do not normally express it. For example, if one expresses a protein from another species or transfects the plasmid into a particular cell type in which that protein is not normally expressed.

Transfecting DNA into cells to induce over-expression of a given protein provides researchers with a window into the role of that protein in the cell and an outstanding opportunity to learn about its function. In addition to introducing DNA coding for the protein itself, by attaching or fusing an

DOI: 10.1201/9781003202974-7

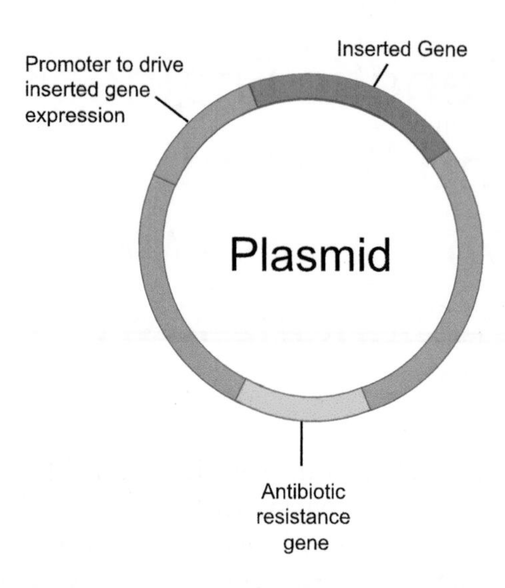

FIGURE 7.1 Schematic diagram of a plasmid used to express a gene in mammalian cells. *Plasmids* are circular-shaped DNA *vectors* used to deliver DNA genes into cells for the purpose of the translation and expression as a protein within the cell. The red region indicates a segment of DNA or gene that is inserted into an area of the plasmid using *restriction enzymes* or enzymes that specifically cut select DNA sequences, thus providing a place for insertion of the desired gene. The *promoter* region is required for transcription of the DNA into RNA, a requirement before translation of the mRNA into protein. Plasmids need to be amplified in bacteria prior to their purification and introduction into eukaryotic host cells; therefore a gene that provides antibiotic resistance is usually included in the plasmid. This allows exclusive selection of bacteria that contain the plasmid as the addition of antibiotics in the media will allow only bacteria containing the plasmid to grow and replicate.

extra piece of adjoining DNA (either at the beginning or end of the gene) that codes for a "tag" and allows labeling of the protein of interest (for example, the jellyfish green fluorescent protein; see Chapter 24), researchers can follow this fusion protein when transfected into cells by microscopy, and determine to which region of the cell the protein localizes, what other proteins and pathways the protein interacts with, and ultimately, the precise function of the protein. The way researchers might obtain clues about the function of such a protein through over-expression experiments may appear counterintuitive. This is because over-expression of a protein will sometimes impede a given biological pathway, likely due to high levels of that protein *saturating out* and interfering with the function of other proteins in the same pathway, thus leading to functional problems in the cell, and hinting at involvement in a given cascade or pathway. However, a more precise way to readily and definitively determine protein function would have to wait until the late 1990s for silencing RNA (siRNA), which will be discussed below.

The advances in molecular biology techniques have also led to dramatic progress in biochemical, chemical, and structural biology, because in addition to enabling the transfection of genes into cells, the new techniques also paved the way for researchers to generate large amounts of any specific protein in bacteria, and then purify that protein to obtain material to study. The use of bacteria as "protein expression factories" has been instrumental for structural biologists and has provided outstanding opportunities to solve protein structures by crystallography or nuclear magnetic resonance techniques, greatly enriching the understanding of protein function at the atomic and molecular level. Why is this important in biomedicine? Among the many reasons is that by knowledge of the precise structure of proteins, some of which are key targets for drugs to treat cancer and other diseases, researchers can better predict which compounds might effectively bind and neutralize these proteins, helping to prevent disease or alleviate symptoms.

To properly understand and unambiguously assign function to select proteins in the cell, a roundabout way such as over-expressing them is obviously not ideal. The best way to evaluate the function of a given protein is to determine what happens (or *what does not happen*) in the *absence* of that protein. The revolution in genetic engineering has led to many ways to knock-out the function of specific proteins, in a variety of cells as well as in whole animals ranging from yeast to mammals (usually mice) and a variety of additional organisms including *Drosophila melanogaster* (the fruit fly), *Caenorhabditis elegans* (the worm), *Danio rerio* (the zebrafish), and others. Until the last 40 years or so, researchers were largely relegated to hoping for good fortune and happening upon cells that either lacked or failed to express a specific protein. Occasionally, such cells could be identified, but as often as not the massive aberrancies in the DNA in these cells made it difficult to really attribute specific function to an individual protein based exclusively on its absence. Another avenue for researchers to pursue was the use of yeast as a genetic model. While yeast cells only have about 6,000 genes compared to the estimated 35,000 in most human cells, as fellow (albeit less complex) eukaryotic cells, yeast cells have a surprising number of proteins with considerable sequence homology to their human homologs and have been extremely useful as a model system. However, as useful as yeast cells have been as a model system for mammalian cells, it was ultimately the development of mouse transgenic mice in the early 1980s, followed by the ability to engineer mouse knock-out models in the late 1980s that has dramatically changed and altered the landscape of biomedical research—all a direct result of the cracking of the genetic code and the accumulating knowledge of protein translation.

MOUSE MODELS FOR STUDYING PROTEIN FUNCTION

The ability to engineer transgenic mice, in which DNA sequences have been inserted into the mouse genome to induce the expression of a select protein, was the culmination of technological advances that built upon the molecular biology revolution of the 1950s, 1960s, and 1970s. It became feasible to use this newly accumulated knowledge to induce gene transfer in mice by microinjecting specific DNA sequences into pronuclei of embryos at the zygote stage. These embryos could then be inserted into female mice, leading to embryonic development and the birth of pups within three weeks. Unlike the initial transgenic mice that were first generated almost a decade earlier, these newer versions of transgenic mice stably carry the new gene and can be bred and subsequently used as model systems for a variety of diseases. Not long afterward, the technology of gene targeting to knock-out mouse genes by homologous recombination opened a much wider window into gene function and added many new forms of mouse disease models. Although extremely useful, these technologies are subject to a significant investment in time. In addition, generating a knock-out mouse raises concerns that the gene of interest might be *compensated* by the function of similar genes in the animal, or just as likely, that the gene of interest might be *lethal* or so important for mouse development that the mice die as embryos and are either not born or die shortly thereafter, and thus impossible to create. Overall, however, the knock-out technologically has served as a gold standard in biomedical sciences to understand gene function and its impact on health and disease.

Mice reportedly share 99% of their genes with humans (Capecchi 1994), and their relatively short breeding time and lifespan (about two years in captivity) combined with their size and relative ease of handling makes them a versatile and practical model for a wide range of diseases. For these reasons transgenic and/or knock-out mice serve as highly useful genetic models, and the former (transgenic animals) are particularly useful when it is known that mutation of a given gene/protein leads to a loss of function in the protein that it encodes. In addition, transgenic and/or knock-out animals serve as outstanding models for the testing of new drugs and treatments. Despite the genetic differences between humans and mice, researchers frequently use mice as the next step for testing the efficacy of new drugs, after initial analyses are first done on cells in culture (Doyle et al. 2012, Hall, Limaye, and Kulkarni 2009).

Although evaluating the total number of knock-out mice generated is a complicated task, in 2017 the International Mouse Phenotyping Consortium began a project to analyze and catalog knock-out models, and found that of the first 3,328 mouse genes analyzed, 360 of them led to disease models, including the first-ever models for several diseases (Meehan et al. 2017). Indeed, as of 2015, the Mouse Genome Informatics resource compiled over 55,000 mouse genotypes that had a phenotypic annotation (some type of observable effect resulting from the altered mouse genes), and about 4,500 genotypes that had disease annotations affiliated with them, highlighting how valuable mouse knock-out models have become for biomedical researchers (Bello, Smith, and Eppig 2015). However, given the time and expense of generating mouse knock-outs, as well as complications in the cases where mouse embryos fail to thrive due to the role of the knocked-out gene product in early development, it was clear that scientists were in need of systems to address the role of specific genes and the proteins that they encode in more simplified cellular systems. Such advances did not become mainstream almost until the twenty-first century with the advent of siRNA and CRISPR technologies.

SILENCE, PLEASE! SILENCING RNA: A NEW METHOD TO STUDY PROTEIN FUNCTION IN CELLS

The first evidence for the existence of an organismal mechanism to prevent gene expression came from studies in petunias in 1990, when researchers were attempting to add exogenous DNA to alter the color of the flowers, and they noticed an unexpected loss in pigmentation (Napoli, Lemieux, and Jorgensen 1990). An additional study done in the worm *C. elegans* suggested that the expression of a specific gene could be blocked by adding small single-stranded RNA that codes for that gene (Guo and Kemphues 1995). However, much of the credit for really cracking open the field of RNA silencing and making it useful to the scientific community—to the point of it becoming a revolutionary technique used routinely by biomedical research labs around the world, as well as a technology to be used for gene therapy—has been attributed to Mello and his colleagues and to Tuschl and his co-workers. In 1998, a paper by Fire et al. from the Mello lab demonstrated that double-stranded RNA could be used in *C. elegans* to specifically inhibit gene expression (Fire et al. 1998), ultimately leading to a Nobel Prize. About three years later, a paper by Elbashir et al. from the Tuschl lab showed that introduction of 21-nucleotide RNA sequences that are matched to a specific mRNA can block the expression of the protein for which that mRNA codes in mammalian cells in culture (Elbashir et al. 2001). Collectively, the use of RNA oligonucleotides to inhibit gene expression in cells became known as silencing RNA, short interfering RNA, or siRNA (note: these techniques collectively belong to what is known as RNA interference or RNAi).

One of the critical findings that led to widespread understanding and use of siRNA was the observation that double-stranded RNA, when introduced into host cells, is cleaved into shorter double-stranded RNA duplexes that associate with a complex of proteins known as the RISC complex. Once this occurs, the double-strand RNA is separated, and the *sense strand* is degraded, whereas the remaining *antisense strand* (which is complementary to the target gene mRNA) directs the RISC complex to the mRNA target which ultimately leads to the degradation of that select mRNA, thus preventing it from undergoing translation into a protein product (Figure 7.2). Over time, as the existing protein in the cell eventually undergoes degradation, and new protein fails to undergo translation due to the siRNA, the level of that specific protein in the cell plummets, allowing researchers to study cell function in its absence in the *knock-down* cells.

Once the mechanisms of siRNA were elucidated and its kinks were worked out, commercialization of the process for research labs quickly followed. Companies set up websites where researchers could essentially choose any gene that they were interested in targeting, and algorithms were formulated to predict RNA duplex sequences that would likely be effective in blocking gene expression of the target protein. siRNA rapidly became the gold standard for researchers to test protein function. Indeed, scientists aiming to publish their research addressing the function of a given protein using cell-based systems over the past 20 years have often been required to demonstrate what happens upon siRNA-based knock-down of that protein, almost as a prerequisite to acceptance by reviewers.

The advances in biomedical research resulting from the technology of siRNA have been mirrored by progress for translational and clinical researchers. There are currently over 20 different clinical trials taking place using various RNA interference strategies (including siRNA, and micro RNA [miRNA], an additional form of inhibitory RNA discovered in 1993 [Lee, Feinbaum, and Ambros 1993]), with many more trials planned. Among the studies underway are the treatment of age-related macular degeneration, a currently untreatable disease that leads to loss of vision, the treatment of hypercholesterolemia by the targeting of apolipoprotein B, treatments for asthma and other respiratory illnesses by knocking down a tyrosine kinase protein known as Syk, and the treatment of a wide variety of solid and other tumors by knocking down proteins that are involved in promoting cell growth.

The use of siRNA in biomedicine clearly highlights the need for fundamental biomedical science; without understanding the process of protein translation derived from the numerous studies done between 1950 and 1970, it would not have been possible to discover and capitalize on the

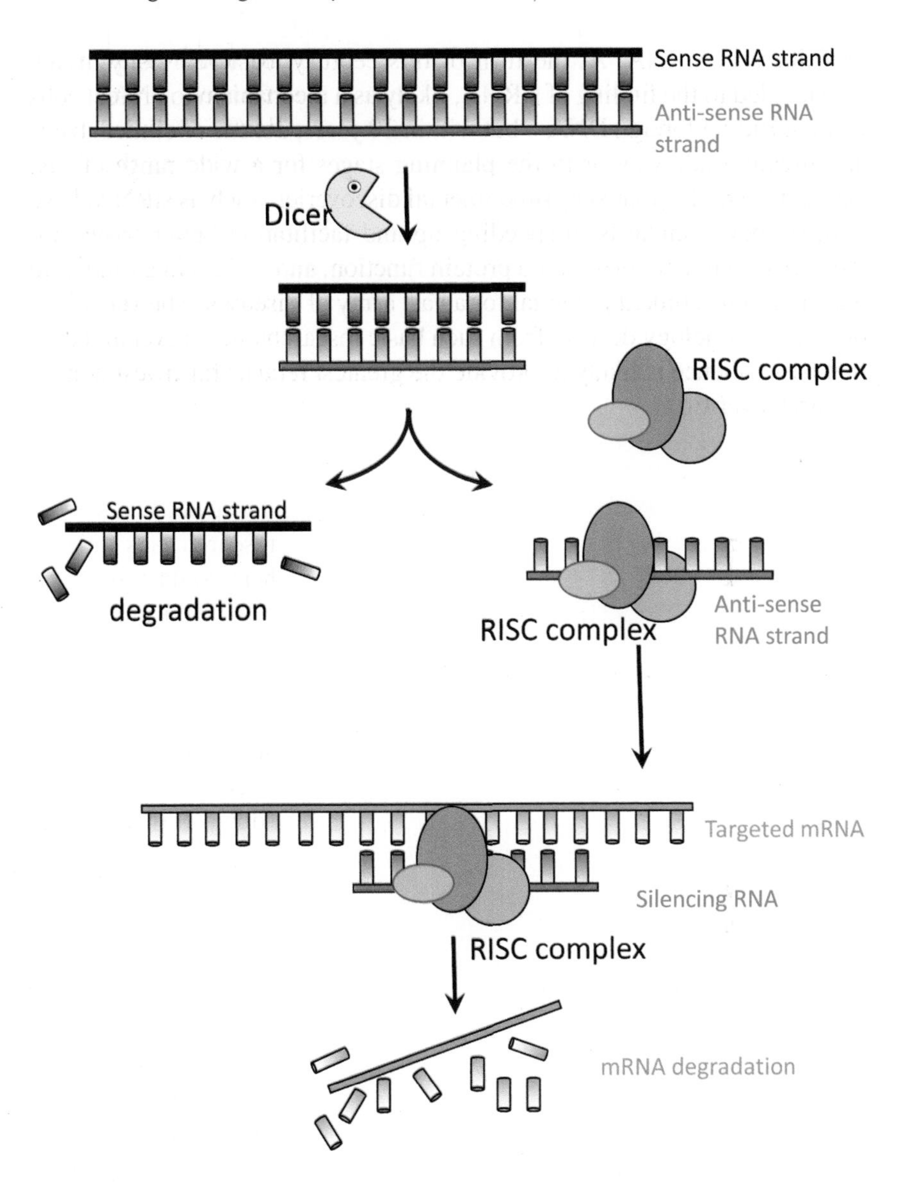

FIGURE 7.2 Silencing RNA (siRNA). A double-stranded RNA molecule is first shortened by the action of an enzyme known as Dicer and then separated into individual *sense* and *antisense RNA strands*. While the sense strand is degraded, the antisense strand is bound by a protein complex known as the RISC complex. Using the antisense strand to specifically identify the target mRNA to be degraded, the RISC complex degrades that mRNA target and thus prevents its translation into protein. By specifically blocking translation of a given new protein, over time that existing protein will undergo degradation and its levels in the cell will be significantly decreased.

usefulness of siRNA. At the turn of this century, basic curiosity-driven research led to the finding of siRNA, likely as a mechanism to protect cells from intruding foreign DNA—but within 20 years, dozens of clinical trials are already underway or in the planning stages for a wide range of diseases. It is notable that very fundamental discoveries, such as siRNA, have both set new standards in speeding up and facilitating better basic biomedical research to understand protein function, and at the same time have had enormous clinical potential for a vast array of diseases. The versatility of a new technology derived from such basic research proves over and over that basic science is likely to provide the greatest returns for investment in research over time.

8 CRISPR, a New Technique for Gene Editing***

The discovery of CRISPR, which stands for clustered regularly interspaced short palindromic repeats, may be one of the most significant biomedical discoveries of our time. But is it really a "discovery"? After decades of painstaking research that has ultimately led to the development of CRISPR, a system that relies on precise and efficient gene editing, the discovery of CRISPR is now culminating in the course of important clinical trials. Indeed, rather than a singular discovery, it is more appropriate to view CRISPR as emerging from an entire series of observations, each of which relied upon previous findings and slowly but surely contributed to the eventual idea that CRISPR could be used for gene-editing in humans. In many ways, the initial fog and uncertainties of CRISPR's origins are not unlike the DNA research that led to the cracking of the genetic code 50–70 years ago. Although perhaps more serendipitous and less targeted (at least initially) than the studies that moved toward uncovering the genetic code, in both cases years of dedicated research from many groups scattered across the world—and not a single "Eureka experiment"—ultimately led to major advances in biomedical science and clinical medicine.

In a recent paper published in the journal *Genetics in Medicine*, Timothy Caulfield and his co-authors cited various publications in the popular press and called CRISPR "one of the century's most important discoveries" noting that CRISPR "is often discussed in terms of its momentous potential impacts" (Marcon et al. 2019). Indeed, two researchers, Jennifer Doudna of the University of California, Berkeley, and Emmanuelle Charpentier of the Max Planck Unit for the Science of Pathogens, in Berlin, Germany, were awarded the 2020 Nobel Prize in Chemistry for their contributions to the development of CRISPR, further highlighting the significance of CRISPR.

What is CRISPR, and why is it so important? In simple terms, CRISPR is really a form of adaptive cellular immune system held by bacteria that endows these microbes with the ability to recognize, target, and attack foreign DNA—in other words, a bacterial defense system to protect against viruses. While there are a variety of different "formats" for CRISPR

DOI: 10.1201/9781003202974-8

defense systems in bacteria, the "Type II" system is perhaps the simplest and the one on which gene editing systems are now based. In general, the CRISPR gene locus (the region where the genes involved in CRISPR are located) contains three parts (Figure 8.1): (1) the CRISPR array, (2) the Cas9 gene along with three additional genes, (3) the TracrRNA. The CRISPR region of the DNA, parts of which were originally identified

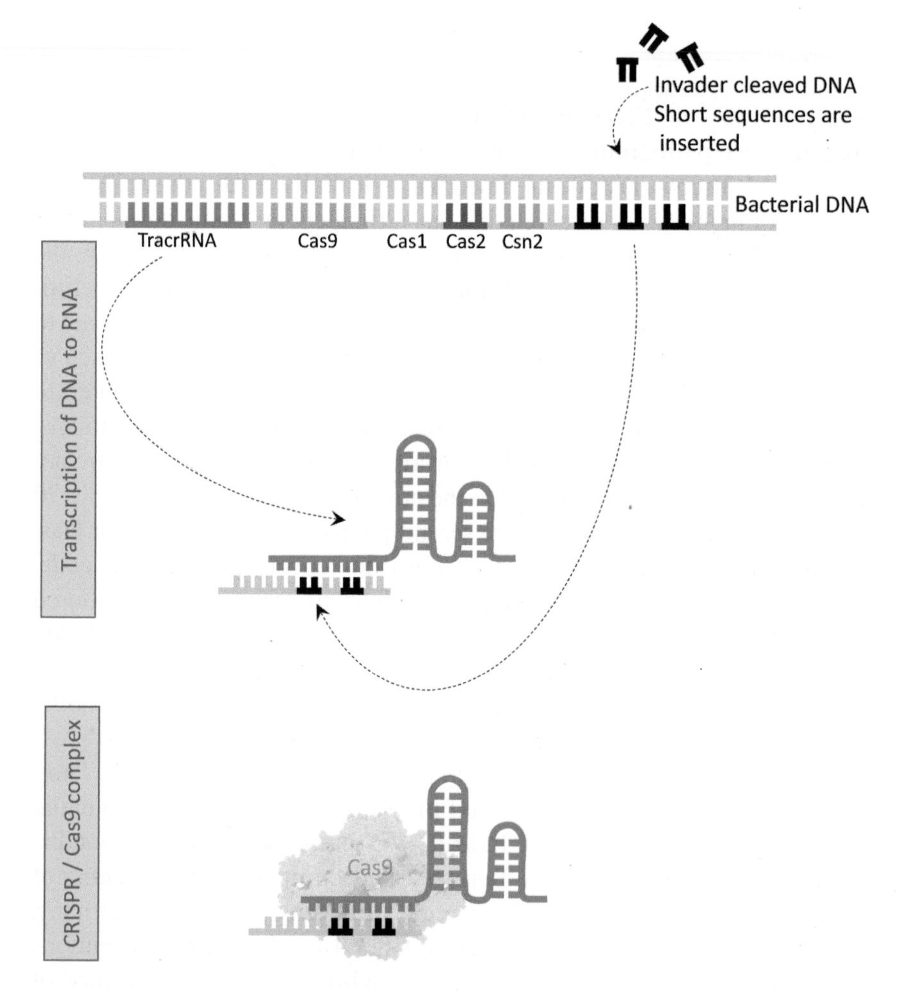

FIGURE 8.1 Arrangement of the CRISPR gene locus. The CRISPR gene locus consists of the TracrRNA (left side), the Cas9 Cas1, Cas2, and Csn2 region (middle), and the CRISPR array (right side). RNA transcribed from the CRISPR array of the bacterial genome providing the selectivity and adaptivity of the immune response is coupled with RNA transcribed from the TracrRNA to form a hybrid RNA molecule. This molecule then is bound to the Cas9 protein that is also generated from mRNA transcribed from the CRISPR gene locus.

and reported by Francisco Mojica of Spain in 1993 (Mojica, Juez, and Rodriguez-Valera 1993), contains the selective sequences that, once they are transcribed to RNA, are used by the bacteria to specifically recognize and bind the DNA of foreign invaders such as viruses and bacteriophages. This specific region, made up of *spacers*, is set between repeated DNA sections within the CRISPR array region. Perhaps one of the most fascinating questions is how do the specific DNA elements that bear homology to the DNA of the invading pathogen become part of the bacterial CRISPR array? This is a key question, because it provides the *adaptive* response for the bacteria—the ability of the bacteria to evolve and be able to attack different and new invading pathogens. To do this, the bacteria take advantage of three of the genes that are sandwiched between the CRISPR array and the TracrRNA. These three genes, called Cas1, Cas2, and Csn2, are involved in cutting up invading pathogen DNA and inserting it into new spacers within the CRISPR array. The fourth gene, Cas9, is ultimately involved in cleaving the DNA from invading pathogens (the DNA that is complementary to that found within the existing spacers). In other words, the three genes Cas1, Cas2, and Csn2 provide the specificity and adaptiveness of the defense system, whereas Cas9 is the workhorse gene that codes for a protein that is brought to the invading pathogen to chew up its DNA and render that pathogen unable to replicate and survive.

In practice, the way that the system works is as follows (Figure 8.2): having already used the Cas and Csn genes to acquire new spacers for the CRISPR array site, that region of the bacterial genome (within the CRISPR locus) is then transcribed from DNA to RNA. At the same time, the TracrRNA is also transcribed, and the two sets of RNA can now pair up through their homologous regions to form a single hybrid RNA molecule. Transcription of the Cas9 and Cas1, Cas2, and Csn2 genes to mRNA also occurs, and these newly transcribed mRNAs are then translated into separate proteins. As noted above, while the Cas1, Cas2, and Csn2 are essentially "scout" proteins to acquire new spacers for the CRISPR array allowing the bacteria to recognize and protect itself from various viruses and bacteriophages, the Cas9 protein forms a complex with the TracrRNA/spacer RNA hybrid. It is then the specific spacer sequences that are used to hunt and seek out the invading pathogen DNA and bring the Cas9 to that foreign DNA for the purpose of its cleavage and inactivation.

The role of the repeat and spacer sequences, coupled with the very enigmatic finding that bacteria actually contain regions of DNA that resemble those of their invading pathogens, took decades to understand. The research was slow, murky, messy, and complex. But it moved forward, as science always does, step by step, and once it was finally understood, it was quickly co-opted as a potential system for the editing of human

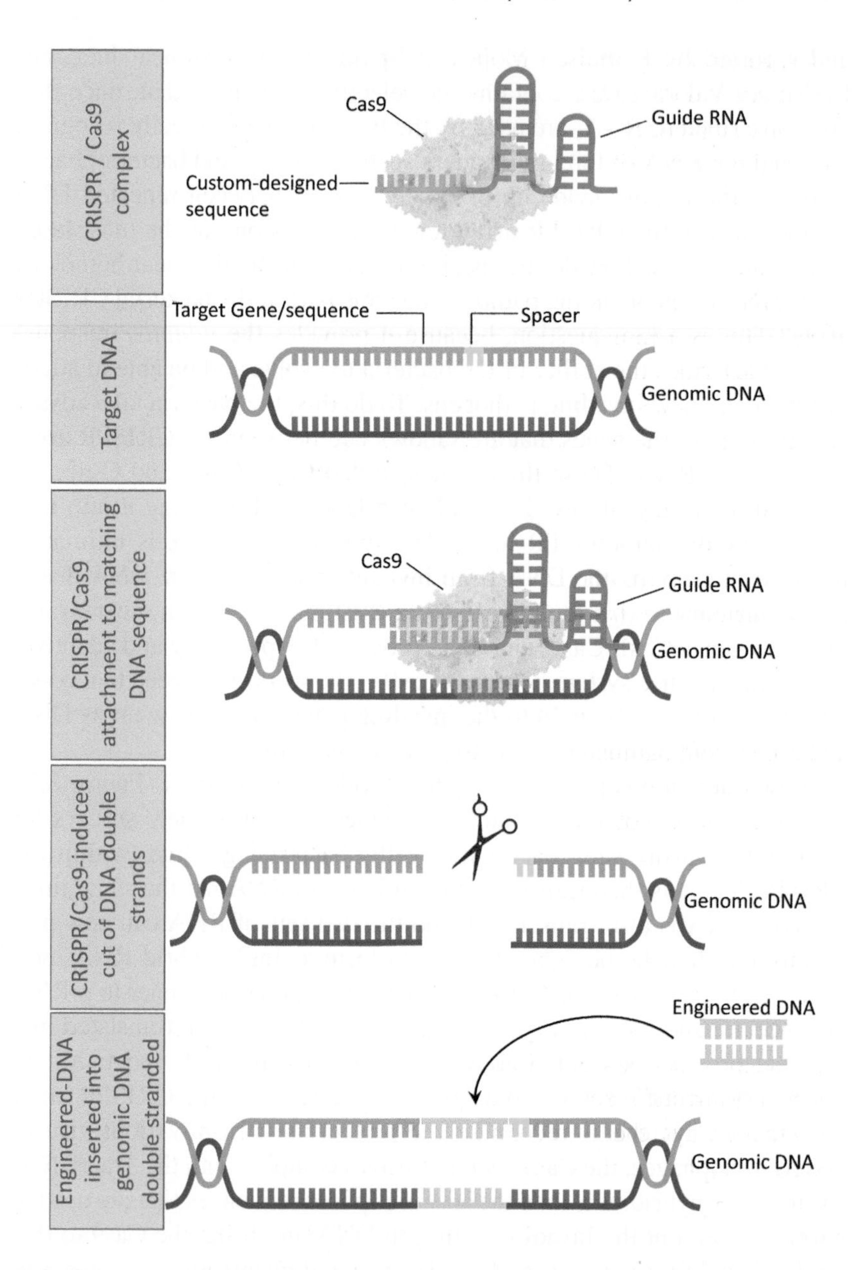

FIGURE 8.2 CRISPR-mediated insertion of DNA into the genome. Guide RNAs serve as custom-designed sequences to mimic the CRISPR locus and bind to the desired target sequence. The attached Cas9 protein cleaves the target sequence at the intended region of the target gene, either neutralizing the gene, or modifying the genomic DNA, potentially by adding or removing parts of an existing gene.

genes. Remarkably, at the start of the CRISPR journey, no one could ever have predicted the immense biomedical implications; after all, why would repetitive sequences—or even parts of invading pathogen sequences in the bacterial genome—be of interest to modern medicine?

In his fascinating article "The Heroes of CRISPR," Eric Lander outlines the long, convoluted history of how the repetitive bacterial DNA sequences discovered by Francisco Mojira in the late 1980s and early 1990s have, over three decades later, led to another major revolution in gene editing along with promising new clinical trials (Lander 2016). During this time, Lander describes the research performed at over a dozen locations across the globe, by researchers coming from vastly different areas of study. As is so often the case in science, the quest for understanding can lead to significant new insights that were never anticipated. Mojira's repetitive sequences were merely a curious phenomenon when he began looking into them. At the time no one anticipated that they would represent a remarkable adaptive immune system that microbes use to defend themselves against invading bacteriophages (the equivalent of a virus that attacks bacteria)—that insight came only after many years of research by multiple labs. Indeed, a significant portion of the research was carried out by labs from the food industry—because at the time it was primarily thought that a bacterial defense system against pathogens would be of importance to those working on fermentation and other processes involving bacteria.

While it was not until well into the twenty-first century that it was proposed and subsequently proven that CRISPR serves as an adaptive immune system to protect bacteria from foreign DNA, it took even longer until researchers understood *how* the system actually functions. Indeed, it was not until the elegant breakthrough studies of Luciano Maraffini and Erik Sontheimer in 2008 that it was clearly demonstrated that the CRISPR system works by cutting DNA, thus leading the researchers to propose that the system might be repurposed for gene editing in a variety of different systems (Marraffini and Sontheimer 2008).

In many ways, the history of the discovery of CRISPR and its mobilization as a tool for gene editing mirrors the great discoveries that ultimately led to cracking of the genetic code. It began with the serendipitous discovery of repetitive bacterial sequences and remained low-key in research circles from the late 1980s until the early 2000s. Following Mojica's lead, the next significant finding in the CRISPR arena was probably that of Alexander Bolotin and his co-workers in France, who discovered the Cas9 gene (Bolotin et al. 2005). This was followed by the studies of Eugene Koonin at the National Institutes of Health in the US, who first proposed that the CRISPR system might serve as a defense system against foreign pathogens (Makarova et al. 2006), which was demonstrated by Phillippe Horvath and his co-workers just a year later (Barrangou et al. 2007). John van der Oost and colleagues in the Netherlands then showed that the spacer sequences in the CRISPR locus that are derived from bacteriophages are transcribed into CRISPR RNA molecules that help to guide the Cas9 to bind and cleave invading foreign DNA sequences (Brouns et al. 2008). While, as noted, Maraffini and Sontheimer showed that CRISPR targets DNA (rather than RNA) (Marraffini and Sontheimer 2008), it was the work of Sylvain Moineau in Quebec City, Canada, that demonstrated that CRISPR-Cas9 specifically targets DNA at positions that are three bases upstream from a select sequence they identified, which is now known as a proto-spacer adjacent motif, or PAM sequence (Garneau et al. 2010). In 2011, Emmanuelle Charpentier, who was at Umea University in northern Sweden and later at the University of Vienna in Austria, identified the TracrRNA that hybridizes with the CRISPR RNA to generate a hybrid RNA molecule that we now know interacts with the Cas9 to target it to the foreign pathogen DNA for cleavage (Deltcheva et al. 2011). While these discoveries, culminating with Charpentier's published work in 2011, cracked the mechanism of CRISPR function in bacteria, the race now began in earnest to take advantage of CRISPR as a potential tool for gene editing in eukaryotic systems.

Given the enormous stakes of patenting a game-changing system for gene editing, one that is currently under numerous clinical trials, it is perhaps not surprising that there are disputes over the harnessing of CRISPR and lawsuits have been filed over patent rights. Virginijus Siksnys and his colleagues at Vilnius University of Lithuania were the first to characterize the mechanism by which Cas9 cleaves DNA (Sapranauskas et al. 2011, Gasiunas et al. 2012), findings that later aided researchers in strategies for adapting CRISPR to gene editing. Charpentier then collaborated with Jennifer Doudna at the University of California Berkeley to show that a single CRISPR RNA/TracrRNA could be fused together to generate a guide to help target Cas9 to a specific DNA sequence for cleavage (Jinek et al. 2012). At the Broad Institute in Massachusetts, Feng Zhang and his colleague George Church at Harvard University, and their research teams made considerable advances by demonstrating that the CRISPR system could be used to do gene editing in eukaryotic cells (Cong et al. 2013, Mali et al. 2013). Without first gaining an overall understanding of how the CRISPR system works, it would not have been possible for these scientists to ingeniously adapt CRISPR for biomedical purposes.

Before discussing the ramifications of CRISPR technologies for individual biomedical applications, it would be remiss not to briefly mention the enormous potential for its use in what have been termed *gene drive* strategies (Esvelt et al. 2014). Indeed, the idea of (for example) genetically engineering sterile organisms or plants incapable of passing on certain traits for release in the wild to mate with endogenous populations, with the goal of preventing the proliferation of pests such as rats or mosquitoes, is becoming increasingly popular. However, the major ethical concerns of permanently altering a species (even a pest) with the long-term effects of such genetic modifications simply not known render this idea a controversial and difficult one for bioethicists and laypersons alike. Given the relative ease with which CRISPR gene editing can be achieved, it is not surprising that such immensely complex ethical questions have arisen, from the unapproved engineering of human embryos/babies so that they (theoretically) are immune to HIV (Cyranoski 2019), to the complex questions as to whether it is ever ethical to use CRISPR gene editing to alter a species.

As of 2020, there were over 20 different clinical trials involving CRISPR/Cas9 gene editing, and at least as many additional trials using other forms of (perhaps less convenient) gene editing, such as those based on transcription activator-like effector nucleases (TALEN) and zinc finger nucleases (ZFN) (Hirakawa et al. 2020). While most of these trials involve *ex vivo* strategies, meaning that DNA is taken from cells derived from the body and the CRISPR engineering is done *in vitro* on the cells prior to returning them to the body, some of the trials are also being done *in vivo*. An example of a disease being targeted with CRISPR/Cas9 technology *in vivo* is Leber congenital amaurosis, an eye disease that impacts the retina. In this case, targeting to the retina is done by injection with the goal of using CRISPR to remove an alternative splice site for the protein CEP290 in retinal cells. Other examples include the use of CRISPR delivered to vaginal cells to induce various genetic modifications to impede human papilloma virus (HPV) infection and prevent cervical cancer. However, among the most exciting uses of CRISPR/Cas9 is its harnessing to impart T cells with enhanced function to attack and neutralize cancer cells. Indeed, the ability to remove immune cells from the blood and apply CRISPR/Cas9 to engineer enhanced T cells in an *ex vivo* setting confers many advantages over the more complicated *in vivo* delivery methods, all of which require sophisticated systems to target the CRISPR/Cas9 to the correct cells in the body. Accordingly, a number of CRISPR/Cas9 studies have attempted to remove T cells from peripheral blood and use CRISPR/Cas9 to neutralize the programmed cell death protein 1 (PD-1) gene, with the goal of impeding this T cell protein that normally restricts the activation of T cells once they bind to their ligand, known as PD-L1. Since high levels of PD-1 on the T cell membrane dampen the activation of these cells and prevent their killing of tumor cells, removal of PD-1 is a general anti-tumor strategy.

One of the most marked advances in gene editing and cancer treatment comes from coupling of CRISPR-based gene editing with chimeric antigen receptor T (CAR-T) cell technologies. This is perhaps one of the most promising of the new adoptive cell transfer (ACT) methods, whereby T cell receptors are selectively engineered to render them capable of binding to and attacking tumor cells. The convenience of CRISPR/Cas9 systems for gene editing makes them an obvious candidate for rapidly advancing ACT and CAR-T technologies, and further illustrates how in the span of less than a generation of research—in the course of a career of a scientist—it is possible to move from the most basic of biomedical discoveries (or even biological discoveries) to the most complex clinical trials designed to treat horrific diseases that until now have been beyond the reach of modern medicine.

9 Connecting Mutations to Disease

*Abnormal Proteins as a Cause of Disease***

During the time of the great revolution in molecular biology and the discoveries pertaining to the mechanisms of protein synthesis, researchers were also making great efforts to understand how specific proteins—or more accurately, mutations in specific proteins rendering them as "aberrant proteins"—could cause disease states. It was in the background of the great advances that determined the role and function of DNA that protein chemist Linus Pauling and his co-workers made huge strides toward demonstrating a key axiom that every biomedical researcher and doctor today takes for granted—that abnormal protein function is the ultimate cause of nearly every (non-infectious) disease known to mankind.

While Pauling was interested, and to some extent even involved, in the race to solve the structure of DNA, ultimately his most significant contribution to science was perhaps establishing a clear connection between aberrant protein and disease. Although in retrospect this may seem to today's scientists a "no brainer," in an era when the relationship between DNA and protein was simply not yet understood, forging a link between a specific dysfunctional protein and a disease was almost a revolutionary idea.

The story of how Pauling became interested in studying sickle cell anemia is somewhat murky, but most versions, including those recounted by Pauling himself, suggest that the idea was initiated following a meeting that he had with a doctor at Boston City Hospital in Massachusetts named William Castle. What is clear is that at some time following his discussion with Castle, Pauling began using the term *molecular disease*, for the first time potentially linking a defective protein with a specific illness.

Pauling learned that in patients with sickle cell anemia, red blood cells (also known as erythrocytes) in the vein go from having a typical rounded button-shape to displaying a scythe-like sickle shape, hence the name. Given the dominant and well-established role of the protein hemoglobin in erythrocyte function by carrying oxygen to the body's tissues, Pauling sagely hypothesized that hemoglobin *is* the cause of the disease, and that

DOI: 10.1201/9781003202974-9

the hemoglobin molecule in sickle cell anemia patients would turn out to have distinct chemical and physical attributes from that of its counterpart in erythrocytes derived from normal individuals.

Linus Pauling has been aptly described as an "ideas person," as one who would lay out dogma-changing ideas, and then recruit outstanding scientists to carry out the scientific laboratory work to support or dispel his ideas (Gormley 2007). Accordingly, having made his prediction his next step was to recruit a researcher to address his hypothesis, and he took a young, recently graduated doctor by the name of Harvey Itano into his lab for a doctoral project: to test whether hemoglobin from sickle cell anemia patients differs from that of normal people without blood disorders.

With direction from his mentor Pauling, Itano worked with great devotion to address the hypothesis that physical, chemical, and/or structural changes in the hemoglobin protein of sickle cell anemia patients were the cause of the aberrant shape of these erythrocytes and the disease itself. As is often the case in science, his initial attempts met with failure. He could find no correlation between the sickle cell shape of the erythrocytes and the types of chemical bonds present in the hemoglobin. In addition, he was unsuccessful at using X-ray diffraction to discern differences in the structure of hemoglobin from patients and healthy individuals. In the meantime, however, Itano did manage to identify a quick and easy way to diagnose sickle cell anemia in the blood of patients (Itano and Pauling 1949). But proving the avant-garde hypothesis that sickle cell anemia is a molecular disease remained an elusive endeavor.

From Darwin's *theory of evolution*, which could not have been formally accepted and understood at a molecular level until nearly a century after its formulation, to the discovery of insulin which first required the development of a simple method to accurately measure blood glucose levels (see Chapter 12), time and time again conceptual scientific advances have relied upon technological advances. In the case of Itano and Pauling's great discovery—that molecular changes in the hemoglobin protein are indeed responsible for sickle cell anemia—it was the requirement for a method to separate proteins based on their charge, known as *electrophoresis*, that saved the day.

Itano and Pauling enlisted the assistance of two additional researchers, Seymour Singer and Ibert Wells, both of whom received the shared credit as co-authors of the famous 1949 *Science* paper along with Pauling and Itano (Pauling, Itano, and et al. 1949), to carry out studies using the still-not-widely-known technique of electrophoresis. Today electrophoresis is one of the most standard techniques used in laboratories to separate proteins and nucleotides in gel matrices most frequently comprised of either polyacrylamide or agarose, respectively. However, in 1949 there were reportedly only several dozen such machines available and in use in the US.

Ultimately, the researchers concluded in their *Science* paper that the charge difference between normal and sickle cell anemia hemoglobin that they observed by electrophoresis was about 0.22 pH units when they measured the *isoelectric point* (the point at which the net charge of a protein is zero, known as the PI)—meaning that the sickle cell anemia hemoglobin would migrate and remain steadfast in a pH gradient at pH 7.09 compared to only pH 6.87 for hemoglobin derived from erythrocytes of healthy individuals (Pauling, Itano, and et al. 1949). Pauling and his colleagues concluded in their discussion that this was likely due to the sickle cell anemia hemoglobin having 2–4 more net positive charges, potentially due to differences in the number of positively charged or negatively charged amino acids in the defective hemoglobin molecule. However, shortly afterward, the authors were forced to retract that conclusion and admit that the differences in the PI of hemoglobin—in other words the pH at which the molecule is neutral within a pH gradient—were not due to changes in positive or negative residues. Instead, Pauling hypothesized that a possible change in the folding of sickle cell anemia hemoglobin might affect the charge of individual residues, thus affecting the overall PI and charge of the molecule.

Whatever the actual reason, Pauling's ingenious idea—one that today helps explain a vast array of genetic diseases—was a breakthrough. Indeed, it was among the first-ever paper to promote the idea that diseases are caused by aberrant proteins—hardly a revolutionary idea in 2020, but it certainly was a novel concept in 1949. But as fate would have it, Pauling and his co-workers were indeed correct with their original publication, and although they had retracted the point stating that differences in positive and/or negative residues in sickle cell hemoglobin resulted in the elevated PI of the protein, their original hypothesis turned out to be correct. In the mid-1950s, Vernon Ingram published several studies that used another new technology called *peptide fingerprinting* to clearly show that peptide fragments from the hemoglobin of patients had a higher (more positive) net charge, and that this was certainly due to the replacement of two negatively charged glutamic acid residues from the normal hemoglobin with neutral valine residues in the hemoglobin of sickle cell anemia patients (Ingram 1956, 1957). In the end, Pauling was entirely right, but validation of his theory again depended on the development of technologies for his vindication. The end result was that researchers and physicians had entered a new era of molecular disease, where it became clear that subtle changes in the amino acid composition of proteins (resulting from gene mutations) are a prime cause of disease.

10 Penicillin

*The Dawn of a New Age of Antibiotics**

In 1929 Alexander Fleming reportedly went on a two-week vacation and accidentally left a Petri plate containing agar that was streaked with *Staphylococcus* bacteria out on his bench. Interestingly, had the plate been kept at cooler temperatures, it is likely that neither the streaked bacteria nor the contaminating mold would have grown. Alternatively, had the plate been kept at warmer temperatures, such as in a 37°C incubator where most bacteria are normally grown, although the bacteria would likely have replicated, the mold would not have thrived at these higher temperatures. However, at the ambient "room temperature," which permits bacteria to grow, albeit more slowly than at 37°C, mold is capable of robust growth. Fleming noticed that the mold, or more accurately, a substance produced by the mold, prevented the growth of certain bacteria (Fleming 1929) (Figure 10.1). Despite the huge ramifications of a substance capable of preventing bacterial growth, for years this potential was not realized because the substance, dubbed penicillin for the name of the mold, was not amenable to being sufficiently concentrated so that it might be used therapeutically. Fleming realized the monumental significance of his serendipitous discovery and he quickly abandoned his original experimental plans for the bacteria and began to focus on the penicillin. Subsequently, it took the biochemical work of the German-Jewish refugee, Ernst Boris Chain, and Howard Walter Florey, a Rhodes scholar who came to the United Kingdom from Australia, to concentrate and purify the penicillin sufficiently so that it could be used in mouse models and then finally to treat people with bacterial infections.

The story of the "rediscovery" of penicillin illustrates three important points. First, akin to many other key scientific findings, serendipity played a key role in the discovery of this antibiotic. It was not planned. Fleming had not set out in a methodical manner to "screen" drugs (as such a method is called today) one after another in an attempt to identify something "medically useful." Instead, he merely carried out his research, and in a rather stunning and unexpected way, casually observed one of the most significant discoveries in biomedical research history. As the great

DOI: 10.1201/9781003202974-10

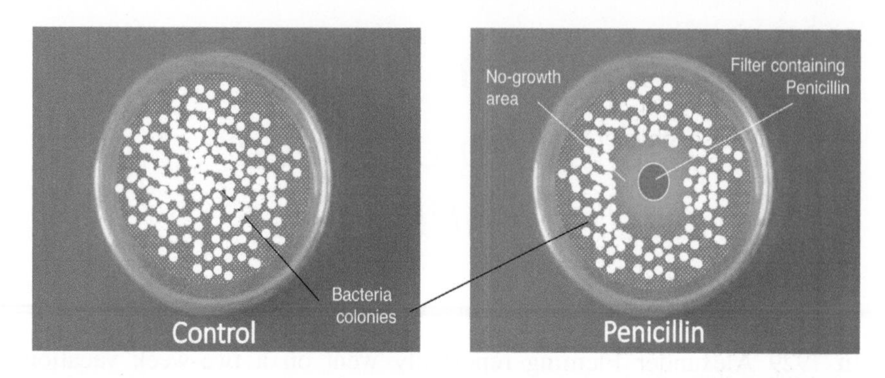

FIGURE 10.1 Penicillin inhibits bacteria growth. Schematic illustration depicting inhibited bacterial growth on a Petri plate when a strip of penicillin is added to the agar on a piece of filter paper (right panel) compared to a control plate lacking penicillin (left panel).

French microbiologist Pasteur is known to have said: "Chance favors the prepared mind." It is noteworthy that in modern translational science, there is a heavy emphasis on the identification of new drugs by a methodology known as *high-throughput screening.* What this means is that scientists have a "library" or collection of drugs and chemicals, often containing drugs that have already been approved for different diseases, in this case known as *drug repurposing.* The scientists then use "brute force of technology" to test, in a fairly unbiased manner, good candidates for treatment of a particular disease, starting first with a series of tests done at the cellular level. This, of course, is an excellent way to rapidly identify new potential treatments. But it is important to point out that understanding the mechanisms by which cells function and thrive offers superb opportunities for researchers to creatively and more selectively design novel treatments and cures, sometimes known as the *candidate approach.* However, the history of science is rife with examples of serendipitous discoveries made by great scientists who despite their careful experimental planning, also kept their eyes and minds open, only to make unexpected new discoveries.

The second important point illustrated by Fleming's discovery and subsequent work on antibiotics is that even the most significant findings, no matter how important they may be, require a foundation of basic knowledge and state-of-the-art technology as well as available methodologies upon which to capitalize. Without the careful step-by-step purifications and advances by Chain and Florey to develop the idea of using penicillin, little progress would have been made toward the widespread use of penicillin in medicine. Fleming was clearly astute enough to realize that the mold was secreting an agent that inhibited bacterial growth. This may sound like a trivial conclusion, given what he observed on the Petri plate, but it is important to realize that although similar results had been reported earlier, a common previous conclusion was that *the mold was competing with the bacteria for oxygen to grow* (Wainwright 1987). Indeed, it is not without irony that Irish physicist John Tyndall made the observation that molds appear to "compete" with bacteria while attempting to disprove the theory that bacteria are "spontaneously generated," and wrongly concluded that this was due to a Darwinian battle between the microorganisms for a supply of oxygen. He did in fact demonstrate that food does not decay under sterile conditions, supporting the findings of Italian physicist Francisco Redi in 1668, who showed that maggots are only observed in decaying meat when there is contact with flies (Levine 2008). It is worth noting that many science historians credit Redi as being one of the first scientists to use bona fide *controls* for his experiments, making him a pioneer in the use of the scientific method for experimentation (Helmenstine 2020). In any case, in a scenario where mold would compete with bacteria for resources such as oxygen, there would be little potential anticipated for the mold to be used therapeutically, leaving open the rediscovery of penicillin to Fleming in 1929. Thus, the proper interpretation and/or usefulness of new discoveries often depends on their timeliness within the overall context of scientific knowledge.

Although there were a series of attempts to use crude penicillin for treating infections early on after 1929, they were met with minimal success. Arthur Dickson Wright, a surgeon at St. Mary's Hospital in London reportedly received filtered yellow fluid from penicillin cultures and attempted to use this on patients (Wainwright 1987). The first authenticated cure of a patient using crude penicillin was by a former student of Fleming, Dr. Cecil George Paine, who used the extract to cure ophthalmia neonatorum in babies (Wainwright 1987). Fleming himself reportedly cured at least one patient, a fellow physician, Dr. K. Rogers, who contracted pneumococcal conjunctivitis in 1932. However, the enormous leap in the therapeutic value of penicillin came from the basic research of Chain and Florey, and key technician Norman Heatley, who cultivated the mold on hospital bedpans, filtered the substance through used parachutes and ultimately managed to extract milligrams of the extract in ether (Wennergren and Lagercrantz 2007).

A third crucial point to appreciate is the significance of a strong foundation of scientific observations throughout history. As noted earlier, science is not performed in a vacuum devoid of previous knowledge and observations. Clearly, Fleming's discovery was really a "rediscovery," since reports back as far as the ancient Egyptians have concluded that throughout history various sorts of molds have been used as salves for infections and to treat skin diseases. In his article "Moulds in Ancient and More

Recent Medicine" (Wainwright 1989), Milton Wainwright also described the use of moldy corn called either kutach bavli or chamka, apparently used by Jews in fourth-century Talmudic times. A paper published in the journal *Science* in 1980 provides evidence that a Nubian bone from the Sudan (from 350–550 ACE) exhibits the fluorescence pattern typically observed for tetracycline-labeled bone (Bassett et al. 1980). This prompted the researchers to speculate that the tetracycline-producing *Streptomycetes* bacteria known to contaminate stored grains may have been ingested to offer this population in the Sudan enhanced immunity to a variety of infections that were common at the time. There are also similar reports from less than 500 years ago, indicating that an English apothecary named John Parkington advocated for the curative use of molds on infections. Many additional anecdotal reports exist of molds having been used to treat infections. Over and over throughout the history of biomedical science and medicine, a recurrent theme is that if the scientific and technological foundation for further developing an important discovery was not yet available, then most likely that discovery would remain an interesting historical anecdote for many years. Eventually, the discovery would finally be "rediscovered" at a time where the fundamental science and technology available would thus allow scientists and doctors to take proper advantage and fully develop the new-found knowledge.

11 Easy to Stomach
The Gutsy Discovery that Helicobacter pylori *Causes Ulcers**

Up until and including the year of 1980, the National Library of Medicine's PubMed site for published papers in biomedical journals (https://pubmed.ncbi.nlm.nih.gov) lists over 10,000 manuscripts with the key words "stomach ulcer." A short meander through some of the latest of these published in 1980—just a few years before the discovery of *Helicobacter pylori* and its involvement in gastric ulcers that eventually led to the awarding of the Nobel Prize to Barry Marshall and Robin Warren in 2005—shows a large number of papers discussing surgical treatments, laser treatments, and the use of cimetidine and other drugs that block acid secretion as treatments for stomach ulcers. Indeed, it was commonly believed that stress and lifestyle, including food intake, were the primary causes for stomach ulcers. The one thing that was not found in the literature was the relationship between the bacterium *Helicobacter pylori* and gastric ulcers.

The discovery of *Helicobacter pylori* as a highly significant cause of many types of gastric and peptic ulcers, as surprising and unexpected as it was in the early 1980s, did not entirely emerge from a vacuum. Although at the time there were relatively few references to the idea that bacteria might play a direct and causative role in ulcers—especially since many doctors and scientists thought that bacteria could not survive in the low-pH environment of the stomach—there were hints and clues in the literature that preceded the findings by Marshall and Warren, who at the time of their discovery worked at the Royal Perth Hospital in Perth, Australia.

First, researchers and doctors noticed that the yeast Candida was frequently observed in surgically resected gastric ulcers (Katzenstein and Maksem 1979). While the authors clearly noted that "Although organisms of the Candida group are probably not directly etiologic in the development of gastric ulcers, it is possible that their presence aggravates and perpetuates gastric ulceration," thus presenting the notion that a microorganism might be related to the disease.

DOI: 10.1201/9781003202974-11

In 1980, around the time that Marshall and Warren began their studies on biopsies from patients with gastric ulcers, standard treatment frequently included the use of drugs that prevented the secretion of gastric juices, such as cimetidine. In one paper published in *The Lancet*, the authors carried out bacteriological studies on the gastric juices obtained from patients to study the range and number of various bacteria found within these secretions (Ruddell et al. 1980). Since the study actually showed increased levels of bacteria following the treatment, it did not exactly support the radical idea that a bacterial infection might be the root cause of many of the gastric ulcers. But in gathering more basic information, this may have ultimately helped Marshall and Warren to formulate their ideas.

In an interview that he gave to *Discover Magazine* in 2010, Barry Marshall explained why the prevailing dogma regarding the causes of gastric ulcers was wrong (Weintraub 2010):

> Eventually doctors realized they could see the ulcers with X-ray machines, but, of course, those machines were in big cities like New York and London—so doctors in those cities started identifying ulcers in urban businessmen who probably smoked a lot of cigarettes and had a high-pressure lifestyle. Later, scientists induced ulcers in rats by putting them in straitjackets and dropping them in ice water. Then they found they could protect the rats from these stress-based ulcers by giving them antacids. They made the connection between ulcers, stress, and acid without any proper double-blind studies, but it fit in with what everybody thought.

When Marshall was in his third year of training in internal medicine, he linked up with pathologist Robin Warren, who informed him that he had seen the same strange S-shaped helical microorganisms in biopsies from numerous patients with ulcers. Cultures showed that these bacteria, later identified as *Helicobacter pylori*, were extremely slow-growing, perhaps helping to explain why they had not previously been fingered as potential causative agents.

The findings of Marshall and Warren, and particularly their conclusions that ulcers are caused by *Helicobacter pylori*, were met with deep skepticism from fellow gastroenterologists and researchers. As Marshall explained: "To gastroenterologists, the concept of a germ causing ulcers was like saying that the Earth is flat." Apparently, a more radical way of convincing colleagues and the general public was needed. Marshall turned to the biomedical version of "put your money where your mouth is"—which can perhaps be rephrased as: "put your treatment where your gastritis is." Marshall collected bacteria from a patient who had gastritis, cultured the bacteria, and used a battery of antibiotics to determine which of these was the most efficient at killing the bacteria *in vitro*. He then used the antibiotics that were successful, bismuth and metronidazole, to successfully treat the patient. Confident that the bacteria had indeed caused the gastritis and that the antibiotics could successfully treat the infection and disease in a patient, Marshall proceeded to drink a broth containing the *Helicobacter pylori* that he had cultured from the patient. This induced inflammation, gastritis, and vomiting, and Marshall monitored his own situation with several endoscopies. Finally, after ten days of suffering (and pressure from his wife who did not want family members to become infected), Marshall took the antibiotics and used his own "gutsy" experiment to demonstrate Koch's postulates and formally prove that *Helicobacter pylori* is the causative agent for gastric ulcers.

While perhaps the discovery of *Helicobacter pylori* as the causative agent in gastric ulcers is not a classic example of basic science leading to clinical gains, it nonetheless has some important lessons for basic biomedical research. The first one is that collecting data, making observations, and forming hypotheses—as done by Robin Warren—are crucial to biomedical advances. The second point is that no matter how astute observations may be, or subsequently, how clever the hypotheses, advances cannot be made on the basis of observations and connections alone—experiments are needed. In the case of discovering *Helicobacter pylori* as the agent responsible for gastric ulcers, there is no doubt that the icing on the cake was provided by the very dedicated and even heroic efforts of Barry Marshall, who while minimizing his risk nevertheless went well beyond the call of duty in showing that Koch's postulates apply.

12 Insulin
*A Hormone Controlling Metabolism**

On the face of it, the discovery of insulin appears to be a very translational-driven project, and science historian Michael Bliss, in his book on *The Discovery of Insulin*, aptly describes how researchers set out to identify the factor that regulates glucose/carbohydrate metabolism (Bliss 1982). While that sounds entirely translationally or clinically oriented, it is clear from Bliss' depiction that major advances were limited until basic scientists worked out accurate methods to measure blood glucose levels, again demonstrating how crucial the advances in basic science and technology are to promote clinical breakthroughs.

The discovery of insulin and its use in the treatment of diabetes came just prior to Fleming's great "rediscovery" of penicillin, in 1922 (see Chapter 10). On the one hand, in contrast to the story of penicillin, the discovery of insulin was not serendipitous, but rather directly targeted. On the other hand, similar to the research leading to penicillin, the discovery of insulin did not emerge from a vacuum. Indeed, for many years, researchers and physicians understood that the pancreas likely was capable of secretions that might be able to alter or regulate carbohydrate metabolism and thus control blood sugar levels. In *The Discovery of Insulin*, Bliss depicts a list of researchers across Europe and in the US who tried unsuccessfully to use pancreatic extracts on severe diabetic patients (Bliss 1982). Finally, intense research at the University of Toronto in Ontario, Canada, carried out primarily by Frederick Banting, Charles Best, John Macleod, and James Collip ultimately identified insulin and paved the way for its use as a first-line treatment for diabetics. Before 1922, the only treatment for those diagnosed with diabetes consisted of a horrific starvation diet for patients, who sometimes received as few as 500 calories a day in an attempt to stem the progress of the disease.

The main driver of the discovery of insulin was physician/surgeon Frederick Banting, who was a shy and not-very-articulate World War One veteran struggling to set up his private clinical practice in London, Ontario. According to Bliss, he had few patients, his rent and debts were piling up, and his childhood sweetheart and fiancé was not pleased with

DOI: 10.1201/9781003202974-12

the situation (Bliss 1982). Banting, tired of his dismal situation, was looking for a change, and found two options: (1) to move to Toronto and carry out some research based on his rather naïve ideas related to pancreatic extracts, and (2) to join an expedition slated to travel to northern Canada as the medical officer for a company. Apparently unable to choose between these two rather radical career changes, Banting flipped a coin, thus deciding to go on the expedition to northern Canada. However, as fate would have it, at the last moment the expedition leader called him to say that they would not be including a medical officer after all, and Banting fell back to "Plan B": work over the summer in the lab of Professor John Macleod at the University of Toronto. In this manner, almost by default, Banting ended up coming for the summer to the lab of Macleod to carry out his research on carbohydrate metabolism.

By all descriptions, Macleod was a very competent and qualified physiologist with broad knowledge of carbohydrate metabolism and physiology in general. Based on records of his early conversations with Banting, he was not impressed by Banting's knowledge of either general physiology or of the current literature regarding pancreatic extracts. Depending on whom one believes, he either cautioned or discouraged Banting, but nonetheless agreed to give him space and resources, as well as a student, Charles Best, to do the work with him. Banting's initial hypothesis was that other researchers had been unsuccessful in obtaining pancreatic extracts with the "internal secretion" that later became known as insulin, because he thought that the "external secretions" (all the gastric enzymes produced by the pancreas) were digesting the insulin, rendering it inactive. His plan, therefore, was to ligate and tie off the ducts that generated the "external secretions," thus allowing those areas of the pancreas to atrophy while leaving the remaining pancreas (and its extracts) more likely to maintain an active and intact "internal secretion" or insulin that would not be taken apart by the external secretions.

The work by Banting and Best was performed at the University of Toronto between the summer and December of 1922, primarily with dogs, rabbits, and later pancreases obtained from slaughterhouses. At some point when there was a modicum of success in obtaining preparations with potential benefit for actual patients, Macleod introduced Banting and Best to John Collip, a biochemist colleague from Alberta who was in Toronto at the time, and his expertise in working out biochemical methods for extraction of the active insulin-containing extracts seems to have further propelled the project forward.

The aftermath of the discovery of insulin is replete with controversy (Rosenfeld 2002). In his book, Bliss highlights scientists who bickered and fought about the scientific credit for their research (Bliss 1982). In truth, assigning credit was no simple matter. Each researcher involved had certain contributions, but for various reasons, it appears that no one person was solely or uniquely responsible for the success. In the end, a contentious Nobel Prize decision was made to award the honor to both Banting and Macleod, whereas Best and Collip were not included. Banting was outraged that Best did not receive his due credit, claiming that "Macleod never did a single experiment." Macleod, on the other hand, being the head of the lab, not only provided the resources, some of the key ideas for successful extraction, and added the biochemist Collip to the team, but he also supported and promoted the research through his talks and help with articles (where he did award equal credit to Banting and Best). Accordingly, while Banting gave half of his award money to Best, Macleod was upset that Collip did not receive his recognition, and he shared half of his prize with the latter researcher.

Why is the discovery of insulin a triumph of basic science? After all, one might argue that in this instance, a clear translational research goal was defined and led to a medical breakthrough. The answer is that basic science in the background of the discovery of insulin is the unsung hero of this story. Why did Banting and his colleagues succeed in Toronto in 1922 whereas 10–20 years earlier, Georg Ludwig Zuelzer and others remained unsuccessful in Europe? The reason, at least in part, is that by 1922, advances in basic science had progressed to allow the development of new assays to accurately and efficiently measure various biochemical parameters, such as the level of glucose in the serum and urine of model animals and humans. Without these significant technological advances in basic research techniques, the discovery of potent pancreatic extracts that contained active insulin would not have been possible. In his book, Michael Bliss clearly suggests that the discovery or "rediscovery" of insulin was ripe in 1922, and that with or without Banting and his colleagues, it was only a question of time until insulin would have been discovered (Bliss 1982).

Many critics of Banting attacked him because his hypothesis that ligating the pancreas would facilitate the isolation of insulin never really led anywhere (and even some of his conclusions in the early experiments were not particularly accurate), and that essentially, the real advance came from the alterations in the preparation of the extracts, ultimately leading to active insulin. It is worth noting that the significance of the biological/biomedical question being addressed is often more important than the specific hypothesis. In fact, often the fact that a hypothesis is proved wrong or irrelevant can be as important as a correct hypothesis for the advancement of science, as long as researchers keep their eyes open and are focused on discovery. As statistician George Box once noted, "All models are wrong, but some are useful" (Box 1979).

13 The Stem Cell
*The Mother of All Cells***

One great discovery that will undoubtedly continue to have major medical ramifications for a wide range of diseases, despite its somewhat controversial (if not also misunderstood) background, is the discovery of stem cells. What is a stem cell and why is it so significant? The human body is made up of a multitude of differentiated cell types and organs, so that retinal cells in the eye differ from cells in the kidney, skin, heart, and brain. At the same time, human embryonic development begins with the generation of a *zygote*, an egg cell fertilized by a single sperm cell. In other words, the single-cell zygote, which leads to the formation of the 150–300 cell *blastocyst*, gives rise to the vast variety of cells, tissues, and organs in the body through the process of *differentiation* (see Figure 13.1). The remarkable concept of the stem cell holds that these special cells somehow maintain their *totipotent* ability to differentiate into any of the multitude of cells needed by the body. Even without applying any special powers of the imagination, it is obvious that the ability to generate, isolate, and/or cultivate such totipotent stem cells would have enormous medical implications; the ability to regenerate neurons for patients with paralysis or neurodegenerative illnesses is just one of the many potential uses that immediately come to mind.

Ironically, the concept of stem cells is almost a mirror image of the problem of cancer; with cancer, researchers want to be able to push undifferentiated or poorly differentiated cells that continue to replicate in an uncontrolled manner into differentiated and more quiescent cells. Stem cell researchers, however, strive to be able to take differentiated cells—for example a skin fibroblast—and return it to its more primordial and totipotent state, where it can then be reprogrammed and then differentiated into any type of cell needed to treat or cure disease.

Tracing the history of the discovery of stem cells is a convoluted process. There are multiple forms of stem cells that differ in their plasticity or potential to generate a wide range of differentiated cell types. For example, embryonic stem cells are the most versatile and can differentiate into any type of cell in the body (see Figure 13.1). Another type of stem cell with major medical ramifications is the hematopoietic stem cell, the precursor of all cells in the blood lineage. Some of the initial studies on stem cells

DOI: 10.1201/9781003202974-13

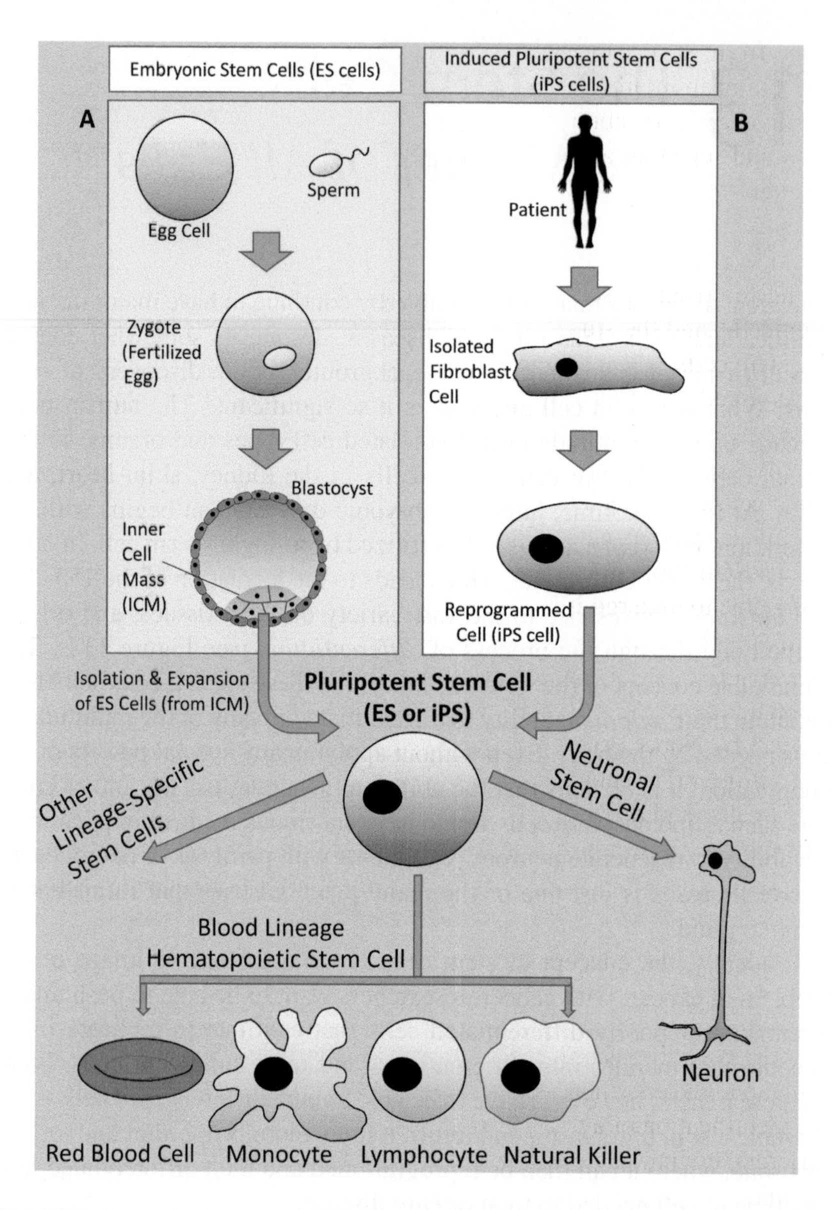

FIGURE 13.1 Embryonic and induced pluripotent stem cells. Schematic illustration showing: (A) embryonic stem cells from the blastocyst have the potential for self-renewal and the ability to differentiate into every cell type in the body, and (B) induced pluripotent stem cells can be derived from a patient's fibroblasts and reprogrammed so that they can differentiate into all the cell types in the body.

came from work with tumors of the sexual organs, known as teratomas and teratocarcinomas (teratos means monster), which were well-publicized due to their very abnormal and monster-like appearance that tended to grow and appear as an array of cells and tissues mixed together, sometimes including bone, teeth, hair, skin, and muscles (Solter 2006). However, many scientists credit two Canadian researchers from the Ontario Cancer Institute, Ernest McCulloch and James Till, with truly bringing stem cells to the forefront of scientific interest and perhaps even with discovering stem cells, and the studies of McCulloch and Till began with hematopoietic stem cells.

As noted, the first type of stem cell identified was the hematopoietic stem cell, a cell with the capability of differentiating into any of the multiple types of cells that comprise the blood, including erythrocytes (red blood cells), first-defender cells such as monocytes/macrophages and neutrophils, as well as natural killer cells and T and B lymphocytes (Becker, Mc, and Till 1963) (see Figure 13.1). Accordingly, a stem cell was defined as being able to undergo infinite self-renewal thus generating more totipotent stem cells, but also capable of differentiating into a variety of cell types (in the case of hematopoietic stem cells, all of the different types of cells of the blood lineage). The first attempted bone marrow transplant was performed in 1958 by Nobel Prize laureate Edward Donnall Thomas, by irradiating and killing off a patient's defective immune system (in other words, the blood cells) and providing the patient with bone marrow from a donor to replenish the immune system. However, this was done before McCulloch and Till had actually identified hematopoietic stem cells (Thomas 2000). Five years *after* the discovery of hematopoietic stem cells, Robert Good performed the first *successful* bone marrow transplant on a 5-month-old male infant whose family had lost no less than 11 family members due to the same severe sex-linked lymphopenic immunodeficiency, using bone marrow from his sister (Gatti et al. 1968).

Despite the enormous medical significance of hematopoietic stem cells, the identification of a totipotent embryonic stem cell that has the ability to differentiate into any cell type was the "Holy Grail" of stem cell research for many years. As discussed earlier, the possibility of creating new neurons for patients with spinal injuries or new insulin-producing pancreatic cells for those with diabetes would be a game-changer for biomedicine. Indeed, in 1981 Martin Evans of Cardiff University in the United Kingdom worked with Matt Kauffman and managed to isolate and grow stem cells from early mouse embryos in culture (Evans and Kaufman 1981). In parallel Gail Martin at the University of California, San Francisco, had similar success (Martin 1981). Evans also was the first to show that such embryonic stem cells from one mouse strain could subsequently be transplanted into the embryo of a different strain, and later implanted into the embryo of a surrogate mouse, ultimately leading to the pups being born each with a mosaic of cells from the different strains. These studies formed the technological basis for the generation of knock-out mice that could be engineered and bred to lack expression of a specific gene (see Chapter 7), and along with Evans, Mario Capecchi of the University of Utah, and Oliver Smithies of the University of North Carolina in the US were awarded the Nobel Prize in 2007.

It was not until almost the turn of the century that *human* embryonic stem cells were first isolated (Thomson et al. 1998, Shamblott et al. 1998). The nearly 20-year lag-time for the discovery of human embryonic stem cells following the initial discovery and isolation of the mouse embryonic stem cells was undoubtedly due to both the technical and complex ethical issues that arose in working with human embryonic cells as opposed to their mouse counterparts. In particular, the ability to obtain fertilized embryos which needed to be destroyed when obtaining the embryonic stem cells became a major ethical hurdle. One of the most significant biomedical advances arose from the stunning discovery by Shinya Yamanaka of Kyoto University in Japan in 2006 who found that by modulating the function of just four genes in adult mouse fibroblasts, he could reprogram the fibroblasts and convert them back into pluripotent stem cells (Takahashi and Yamanaka 2006). Shortly thereafter, both Yamanaka and James Thomson, at the University of Wisconsin at Madison, demonstrated that differentiated human cells could also be "reversed" into pluripotent stem cells, thus bypassing the need for embryos (Takahashi et al. 2007, Yu et al. 2007). Yamanaka was awarded the 2012 Nobel Prize together with John Gurdon of the University of Cambridge in the United Kingdom. Gurdon was credited with being a pioneer in "reproductive cloning" back in the 1960s and having been the first to show that differentiated cells have the potential to revert back to a more primitive state, when he succeeded in taking a frog egg cell and removing the nucleus, and then instead inserting the nucleus from a tadpole intestinal cell (Gurdon 1962b, c, a). The growth of a new frog thus proved that the adult cell still contained all of the information required for the generation of all of the frog cells and was a precursor to Yamanaka's findings almost 50 years later.

The discovery that adult cells could be induced into becoming pluripotent stem cells (known as iPS) marked a revolution in the stem cell field for several reasons. The very complex technical and ethical issues arising from the necessity to derive embryonic stem cells from fertilized embryos (that were destroyed in the process) could be altogether avoided by reprogramming adult cells to become iPS. Moreover, the ability to readily culture adult cells such as skin fibroblasts from an individual to specifically generate muscle tissue or neurons that would be immunologically compatible was just as significant. Indeed, of the cells tested for use in conversion to iPS cells, fibroblasts appear to be among the best suited cells.

Aside from the many approved stem cell treatments using stem cells derived from umbilical cord blood for blood diseases such as leukemias, lymphomas, and a long list of other blood cell proliferation disorders, there is a rapidly growing number of clinical trials underway with iPS cells. While few treatments have been approved thus far, as of 2020 there were several dozen trials involving both embryonic stem cells and iPS cells, for a wide variety of diseases. These include eye diseases such as macular degeneration, as well as β-thalassemia, amyotrophic lateral sclerosis, Parkinson's disease, ischemic heart disease and heart failure, treatment of solid tumors, meniscus injuries, and many other diseases and disorders, and a great deal of promise is anticipated (Zakrzewski et al. 2019, Deinsberger, Reisinger, and Weber 2020).

The development of stem cell treatments is likely to be of major consequence for medicine in the twenty-first century. However, the history of the discoveries of the various types of stem cells that led to stem cell clinical trials and the new field of regenerative medicine highlight the need for basic biomedical research and an understanding of the basic biology. The many decades of research that preceded the big advances and Nobel awards highlight how big leaps often come from painstakingly slow science. Indeed, the combined Nobel Prizes in 2012 given to John Gurdon and Shinya Yamanaka perhaps best exemplify how the most basic biological discoveries in the 1960s by Gurdon ultimately fueled the groundbreaking developments led by Yamanaka in the early 2000s. It took about 45 years, but Gurdon's demonstration of "reproductive cloning" could finally be explained by Yamanaka at the molecular level, and these fundamental insights spurred the development of potential new therapies.

14 Antibodies

*A New Way to Harness the Immune Response***

The discovery of antibodies perhaps best illustrates how scientific advances can be achieved in a variety of ways. Until now, most of the chapters in this book have highlighted the manner in which basic discoveries are made; mechanisms are understood, and in the vast majority of cases, only years later the development of drugs and treatments is facilitated. As the history of the discovery of antibodies demonstrates, this is clearly not the only scenario by which science effects advances. Indeed, as this chapter will show, treatments and medical benefits long preceded the actual discovery and mechanistic understanding of antibody structure and function. However, as this chapter will also illustrate, once the nature of antibodies had been extensively studied and understood, these findings sparked additional new discoveries and vastly significant new medical advances. Ultimately, although mankind was already benefiting from antibodies long before they were even given a name, the scientific discovery of antibodies years later did portend crucial new biomedical advances and treatments for diseases.

In the late 1700s, Edward Jenner was a country doctor in the United Kingdom who performed a highly unethical, but extremely successful experiment that radically altered medicine, public health, and science, and may well have been the first experimental demonstration of the immense power of antibodies. While Jenner had no idea of the concept of antibodies, he was a naturalist and proficient doctor. He knew that the English farm women and "milk maids" who worked with cows and frequently developed wart-like lesions or "cowpox" on their hands rarely if ever contracted smallpox, even though the disease was rampant during this time (Stern and Markel 2005). Reasoning that exposure to cowpox somehow provided a form of protection against smallpox, Jenner decided to inoculate an 8-year-old boy with material from cowpox lesions taken from a milkmaid named Sarah Nelmes. The boy, James Phipps, was the son of his gardener, and Jenner aimed to determine if indeed he became *immune* to smallpox. His experimental design was simple—he extracted fluid from cowpox lesions and injected it at two sites on the boy's arm. Several weeks later he *challenged* the boy by directly exposing him to smallpox, and fortunately he

DOI: 10.1201/9781003202974-14

found that the boy remained healthy. Indeed, he subsequently challenged Phipps about 20 separate times over the coming months and years with smallpox, only to find that Phipps remained healthy and protected from the disease. It is noteworthy that anecdotal stories hold that even before Jenner vaccinations to smallpox were being used. Chinese emperors in the seventeenth century were reportedly inoculating their own children, having learned that protection to smallpox could be obtained by grinding up smallpox lesions from sick people, and blowing the dust into the noses of the *vaccine* recipients, and reports from India indicated that similar vaccinations being given even earlier in the 1500s (Boylston 2012). Indeed, there are even reports of Buddhist or Taoist monks using some form of vaccination well over a thousand years ago (Boylston 2012). However, Jenner's extensive documentation by writing and self-publishing his experimentation, at his own expense, secured him a place in history as the father of modern vaccinations, however unethical and unorthodox his methodologies may have been (Jenner 1798). Strikingly, Jenner's legacy remains in our daily vocabulary, as the word vaccine comes from the Latin *vacca*, or cow, owing to the preventative effect of cowpox inoculation in protecting against smallpox infections.

Building on Jenner's ideas of inducing protection by vaccinating against infectious diseases, in the 1890s in Berlin, Germany, Emil Von Behring (who was awarded the Nobel Prize in 1901) and Shibasaburo Kitasato went a step farther than Jenner to hypothesize and then show that the serum of an animal injected with diphtheria or tetanus could be transferred to other animals and used to protect them from contracting the disease (von Behring and Kitasato 1890). The notion that a protective substance in the serum, coined as *anti-bodies* and later identified as immunoglobulin proteins, could be used to explain Jenner's success in smallpox vaccinations became widespread in the scientific and medical communities. Indeed, the idea of protective serum that can be transferred from person to person, also known today as convalescent serum, is still being used in 2020 to treat diseases such as Ebola and COVID-19, at least experimentally.

Paul Ehrlich (Nobel Prize winner in 1908), also in Berlin, was critical to further advancing the discoveries of Von Behring and Kitasato, and he played an essential role in developing high-quality anti-diphtheria serum for the treatment of infected people. He also found that laboratory animals who were fed low doses of toxins obtained *immunity* to lethal doses of the same toxin. However, at that time the cellular and molecular mechanisms by which such immunity could be attained were not understood. In 1897, Ehrlich postulated a remarkable theory that he termed the "Side Chain Model of Immunity" (Silverstein 2000, Valent et al. 2016). Although this theory turned out to be incorrect, it nonetheless provided crucial insight and ultimately led researchers to understand the mechanisms of immunity, and of course, the nature of antibodies. In this model, Ehrlich hypothesized that immune cells (lymphocytes) expressed multiple types of *side chains* (later these would become known as *receptors*) on the surface of cells that were responsible for binding and allowing different nutrients to access the cell. In his view, toxins from bacteria and other sources could mimic these nutrients and attach to the side chains, thus preventing them from carrying out their essential function in supplying the needed nutrients to the cells. As a result, the cells would respond and generate large concentrations of these very same side chains that would ultimately be released from the cells into the serum, where they might act as *antitoxins* and bind to the toxins and neutralize them before they have an opportunity to harm the cells. While some of the key details are obviously wrong, it is important to note that Ehrlich actually did essentially identify some of the key principles of adaptive immunity in this early hypothesis. In effect, the release of antitoxins into the bloodstream by immune cells in response to a bacterial infection or immunological challenge is remarkably similar to the way in which B lymphocytes and plasma cells generate antibodies that are secreted to the serum. Although Ehrlich's notion that these side chain antitoxin/antibodies are required for providing the cells with nutrients was wrong, the *side chain model of immunity* was a pivotal idea that spurred on immense new understanding in the field of immunology. It is also important to note again that such initial models in science are often incorrect or inaccurate when first proposed, but they provide a new framework for experimentation and enhanced understanding. Ehrlich himself understood the role of models in science, and he adapted and revised his ideas as new data emerged, precisely the way scientists should. As a result, in 1900 he proposed the term *receptor* to replace his *receptive side chains* and his ideas that receptors bind to specific agonists and antagonists have been a foundation of modern pharmacology (Valent et al. 2016).

Despite the emerging significance of immunity, it was another 50 years or so until Mogens Bjørneboe and Harald Gormsen at the State Serum Institute in Copenhagen, Denmark, showed that repeated vaccinations of rabbits led to both a massive proliferation of plasma cells and high antibody concentrations in the serum (Bjørneboe, Gormsen, and Lundquist 1947), consistent with previous observations that patients with higher serum γ-globulin concentrations also had greater numbers of plasma cells. However, it was not until the definitive studies of Astrid Fagreus of the Karolinska Institute in Sweden that it was formally proved that a select type of B lymphocyte known as a plasma cell is *the* cell responsible for the generation and secretion of antibodies to the serum (Fagraeus 1948). In her research, Fagreus was able to unambiguously demonstrate that spleen cultures from immunized rabbits contained plasma cells that secreted antibodies with the capability to bind and neutralize the foreign substance or *immunogen* (Fagraeus 1948). But at the time the source of these remarkable plasma cells prior to their differentiation was not known.

It is important to note that γ-globulin, also known as immunoglobulin G (IgG), is not the only type of antibody, although it may be the most common and arguably the important isotype found in the serum. As it turns out, IgM is the predominant type of antibody that is observed immediately upon infection, and high IgM levels to a specific pathogen typically indicate an active infection by that pathogen. On the other hand, after an *isotype switch* to IgG, these antibodies are then considered to be protective in the serum. Of the remaining isotypes, IgD, IgA, and IgE, the latter is especially important in allergy, and its discovery, while beyond the scope of this chapter, is an intriguing one in its own right (Ribatti 2016).

At the time that researchers were studying γ-globulin in the serum, not only was the source of the mysterious plasma cells not known, but even more importantly, the nature of the immune response and the massive diversity by which antibodies are capable of responding to practically every type of microbial threat was also not at all understood. Danish immunologist Niles Jerne (who was awarded the Nobel Prize in 1984) proposed his "Natural Selection Theory of Antibody Formation" in 1955 in which he favored the idea that there is a vastly diverse range of "pre-made" antibodies circulating in the serum, and when any one of these antibodies encounters a specific pathogen or antigen in the serum, there is a process of "natural selection." This means that the immune system hones in on the specific threat by selectively generating those effective antibodies that bind and neutralize the invader (Jerne 1955). Shortly thereafter, David Talmadge at the University of Chicago published his theory of "Cell Selection" in which he proposed that the diversity of antibodies comes from the level of the individual cell (Talmage 1957). Almost in parallel, Frank Burnet (Nobel Prize, 1960) at the Walter and Eliza Hall Institute in the University of Melbourne, Australia, coined his "Clonal Selection Theory" in which he predicted that each cell generates antibodies of a single specificity, and he speculated that these cells would proliferate in response to binding to a specific antigen. As a result, the concentration of cells that produce an antibody to a specific antigen threat would increase dramatically upon encountering that antigen in the serum and thus increase antibody abundance, resulting in "clonal selection" (Burnet 1957). Needless to say, the notion that each plasma cell generates only a single type of antibody that recognizes a single specific antigen was a breakthrough that would have immense consequences for the future of immunology, and eventually lead to the generation of monoclonal antibodies and novel treatments for multiple diseases.

Despite their brilliance, however, Jerne, Talmage, and Burnet had articulated revolutionary theories, but had not proved them experimentally. Burnet's former student, Gus Nossal, together with Joshua Lederberg (who was awarded the Nobel Prize for work on bacterial mating in 1958) at the Walter and Eliza Hall Institute, clearly demonstrated that single cells will generate only a single type of antibody specific to a single antigen (Nossal and Lederberg 1958), proving *clonal selection theory* beyond a doubt. Thus prior to 1960, scientists understood that antibodies were generated by B lymphocytes that differentiate into antibody-secreting plasma cells, and that each such cell is capable of generating a single type of antibody that recognizes a single antigen (or foreign particle). These remarkable characteristics of antibodies would later form the basis for the incredibly important technological innovation known as *monoclonal antibodies*—the generation of a completely homogeneous population of antibodies that binds exclusively to a single antigen. The enormous significance of this technology both for biomedical research and for the treatment of numerous diseases is impossible to exaggerate.

The Y-shaped symbol of the antibody has become iconic over the last half century, but discovering the structure of the antibody was the "Holy Grail" of a great number of scientists for many years. Rapid advances came from the biochemical approaches of Rodney Porter (University of Oxford, United Kingdom) and Gerald Edelman (Rockefeller University, New York) who were the first to use chemical reducing agents that break cysteine-cysteine disulfide bonds between proteins (or break the disulfide bonds between subunits of the same protein, or even between cysteine residues on the same subunit of a protein). Using such these techniques, these two independent researchers, who each received a Nobel Prize in 1972, learned that a single immunoglobulin of the type known as γ-globulin (IgG antibody) is comprised of two identical heavy chains (50–60 kilodaltons each, and attached to each other by cysteine disulfide bonds), and two light chains (~25 kilodaltons each, and attached by cysteine disulfide bonds to the heavy chains) (Figure 14.1). Together, these chains make up the "Y-like" shape of the immunoglobulin, and the "V-like" region of the antibody (known as the Fab region) has the variability that allows interaction with the vast array of foreign antigens that are encountered within the bloodstream. Through their remarkable crystal structures, Edelman and Porter also discovered the function of the more invariable region of the antibody, the Fc region (constant region). The Fc region is crucial for attachment to effector cells of the immune system; this in turn mediates antibody-dependent cell cytotoxicity (ADCC) of infected cells that

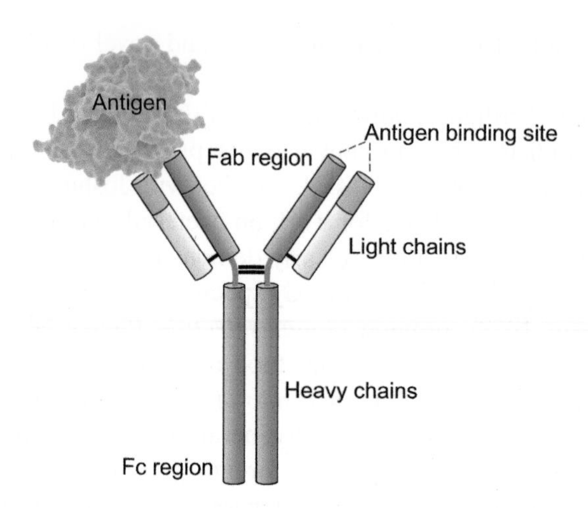

FIGURE 14.1 Schematic diagram of an antibody. The immunoglobulin heavy chains and light chains are each attached to one another by disulfide bonds on cysteine residues (black lines). The antigen-binding variable regions are localized to the tips of the "Y-like" ends of the antibody in the region known as the Fab region and are comprised of both light and heavy chains. The Fc tail (constant region) of the antibody is recognized by specific receptors involved in antibody-dependent cell-mediated cytotoxicity.

are recognized by the Fab regions of the same antibodies, thus linking the effector (killer) cells to the target cells displaying the foreign antigen (Edelman 1973, Porter 1967) (Figure 14.2). In this sense, the antibody serves as a bridge between the foreign antigen and the effector/killer cells that destroy the immunogen or pathogen.

While the mystery of the structure, function, and almost infinite diversity of antibodies slowly began to clarify, the molecular mechanisms by which these remarkable proteins are capable of acquiring their diversity and ability to interact with and neutralize nearly every pathogen known to humankind remained enigmatic until the 1970s. At the time, there were differences of opinion as to whether antibodies were coded by different genes (known as the *germline theory*)—a notion that would require an enormous genome—or whether the variations in the antibody binding regions resulted from changes that occurred during cell replication (known as the *somatic theory*).

In 1974 Susumu Tonegawa (who was awarded the Nobel Prize in 1987) and his colleagues at the Basel Institute for Immunology in Switzerland provided key new data that essentially demonstrated that there are not enough genes within the human genome to account for the immense and nearly infinite degree of diversity of antibodies, thus supporting the somatic

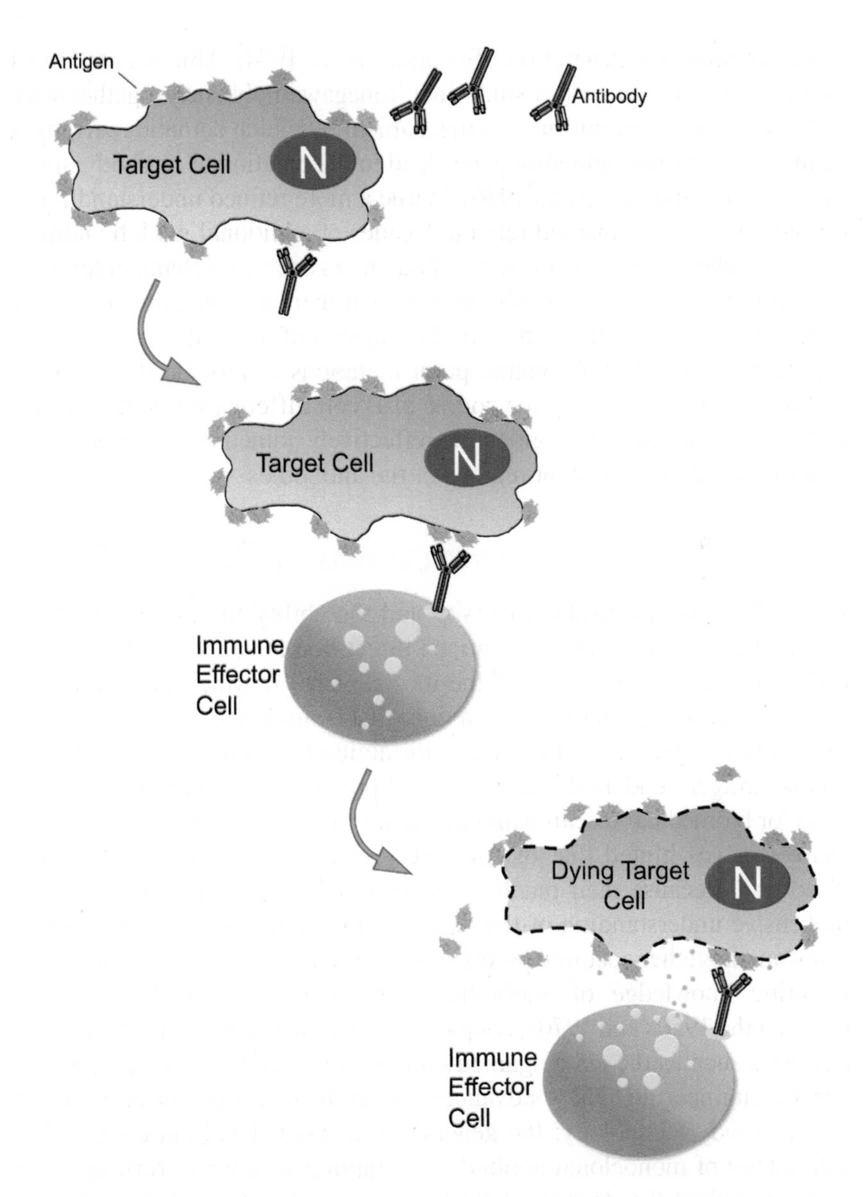

FIGURE 14.2 Antibody-dependent cell cytotoxicity. Free antibodies secreted by plasma cells into the serum bind to an antigen on the surface of a target cell (or on the surface of a pathogen). The Fc regions of the binding antibodies are then in turn bound by receptors on the surface of immune effector cells, which subsequently release proteins that permeabilize the target cell surface membrane and cause the target cell to die.

theory of antibody generation (Tonegawa et al. 1974). This was followed a couple of years later by a study that Tonegawa published together with Nobomichi Hozumi, outlining a mechanism by which somatic rearrangements of the immunoglobulin genes lead to the creation of antibody diversity (Hozumi and Tonegawa 1976). While a more refined understanding of this complex process would rely on decades of additional work by numerous researchers, Tonegawa's studies laid out a general mechanism for antibody diversity. Tonegawa's solution was that there are separate genes that code for the expression of the variable regions of antibodies where there is a high propensity for somatic point mutations to provide the diversity within the Fab, and during the course of B cell differentiation these genes coding for the variable regions are effectively joined together with the genes that code for constant regions of the antibodies.

THE GENERATION OF MONOCLONAL ANTIBODIES

Antibodies have incredible diversity and the ability to bind and neutralize practically every potential antigen threat, foreign particle, or invading pathogen. An understanding of the mechanisms by which these antibodies are generated, along with the axiom that a single B cell or plasma cell generates a single type of highly specific antibody that recognizes only one unique antigen renders these specialized proteins as incredibly important tools for biomedical research and medicine. While at the beginning of this chapter it was hinted that the discovery of antibodies was an "atypical discovery" because their practical medical use actually preceded a comprehensive understanding of the nature of the antibodies themselves, this is not an altogether accurate premise. The reason is related to the enhanced scientific knowledge of antibodies that rapidly developed, especially between the 1950s and 1970s, coupled and intertwined with other scientific advances such as the newly gained understanding of the genetic code and genetic engineering. These advances fueled the development of a crucial new antibody technology: the generation of monoclonal antibodies. The technology of monoclonal antibody generation was a major turning point in biomedicine; this technique takes supreme advantage of the potential of antibodies and has revolutionized both scientific inquiry and medical treatment to this day. First demonstrated by Cesar Milstein and Georges Kohler at Cambridge University in the United Kingdom (Nobel Prize winners in 1984), the generation of monoclonal antibodies has become one of the most important scientific and medical advances in the last century (Kohler and Milstein 1975).

The idea of harnessing the supreme specificity of the immune system to derive useful antibodies was not a new one. Indeed, even today researchers immunize laboratory animals such as rabbits with a specific antigen, often followed by several additional injections to further boost the immune response, allowing them to obtain serum with reactivity to the specific antigen to which the rabbits had been immunized. The antibodies generated by injecting rabbits in this manner against a specific protein target are known as *polyclonal antibodies*, and they are representative of the type of immune response humans have when subjected to a foreign invading pathogen, or for that matter, upon receiving a vaccination. When injected into rabbits an antigen can stimulate a variety of B cells to differentiate and secrete antibodies that show some reactivity to the injected antigen. Accordingly, each B cell clone will differ, and once each B cell differentiates to a plasma cell that secretes a different antibody, a variety of different polyclonal antibodies will be generated and secreted into the blood serum. Indeed, for rabbits (as well as humans) who encounter a pathogen, this diverse and polyclonal response is typically a good thing, because it increases the chances that one or more of the different antibodies will be able to neutralize the invader and collectively the diverse antibodies will overcome the infection. Of course, within the serum there will also be many secreted antibodies that react with any other of the myriad antigens that the rabbit has been challenged with during the course of its lifetime. While this type of polyclonal antibody continues to be incredibly useful in biomedical research, for some research applications the availability of an antibody with singular specificity, derived from a single clone of cells—the *monoclonal antibody*—can be more powerful than polyclonal antibodies. However, perhaps the most significant breakthroughs that monoclonal antibodies have provided are on the medical side. Monoclonal antibodies are increasingly being used for the treatment of a variety of blood cancers such as lymphomas and for some types of solid tumors, as well as a growing number of other illnesses that include psoriasis, macular degeneration, sickle cell disease, osteoporosis, HIV infection, and many other diseases.

As noted, Kohler and Milstein's landmark paper published in 1975, depicting a relatively simple method to generate a clone of cells that secretes antibodies with singular specificity, has changed the face of scientific discovery in virtually every field imaginable (Kohler and Milstein 1975). At the same time, monoclonal antibody technology continues to facilitate crucial new treatments for cancers and a host of other illnesses. The breakthrough was, in retrospect, quite a simple one. These scientists inoculated mice with an antigen (sheep red blood cells in their case, which contain proteins that are "foreign" to mice), and then removed the spleens from these mice because the spleen is the main source of the B cells (Figure 14.3). Since B cells from spleens can be cultured but do not continue to proliferate indefinitely *in vitro* (outside the mouse's body), they used a technical trick and *fused* all of the millions of B cells from the spleen, each one generating a different antibody with its own unique

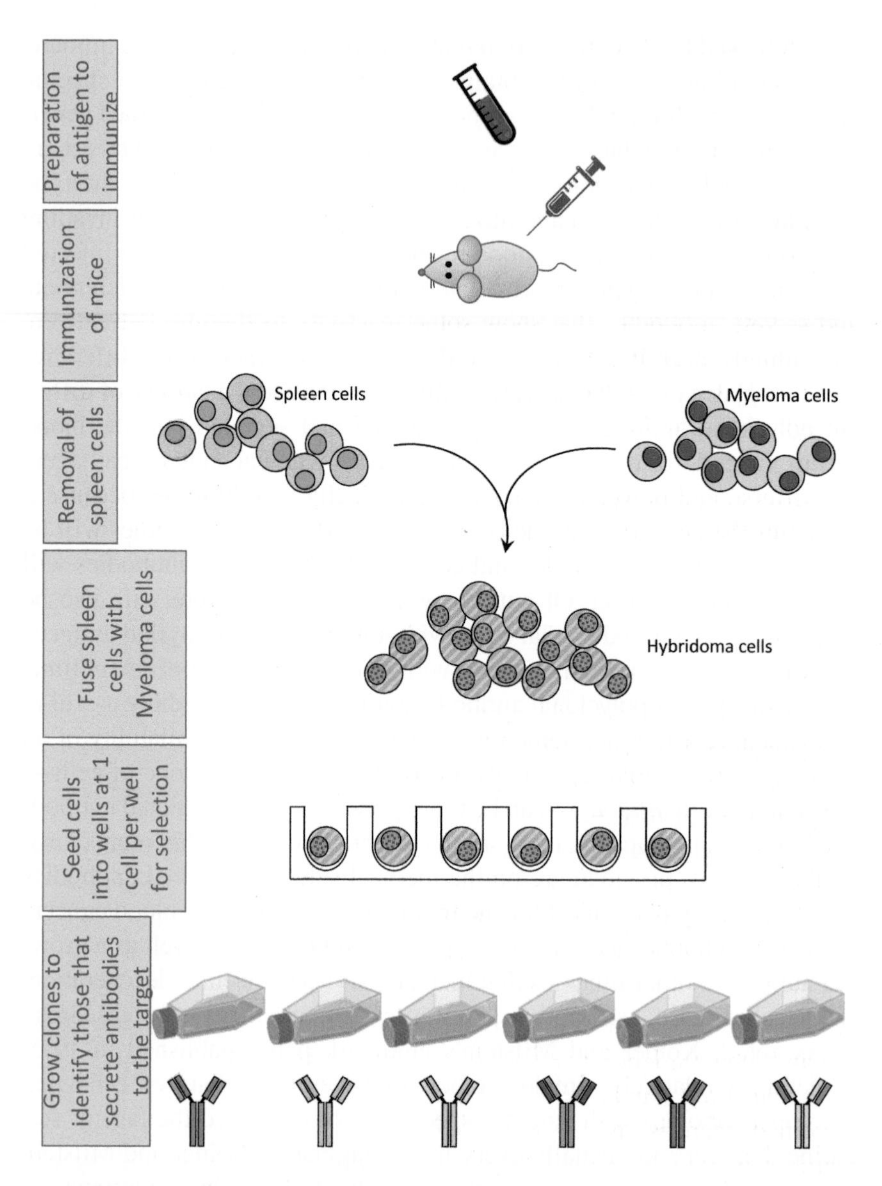

FIGURE 14.3 Monoclonal antibody generation. Mice are inoculated with an antigen. Once an immune response has occurred, the mouse spleens are removed and B lymphocyte cell clones that react to the specific antigen are fused with myeloma cells to generate fused hybridoma cells. The hybridoma cells are then plated at a concentration of one cell per well (or fewer) and cultured to generate a monoclonal population of hybridoma cells that secrete a singular antibody to a specific antigen. The supernatant media from the wells are tested for efficacy in reacting to the specific antigen, and thus hybridoma cells are *screened* to select a monoclonal population that generates an antibody that recognizes the immunogen of choice.

specificity, with a type of transformed myeloma cell that proliferates indefinitely in culture. Having fused the millions of different B cells from the mouse spleen with the myeloma cells to generate *hybridoma cells* that can now grow and continue to produce antibodies, Kohler and Milstein then took the whole batch of fused cells and seeded them on plates with tiny wells at a concentration so that statistically there would be only one cell (or fewer, i.e., none) per well. The single cells in their individual wells were then allowed to proliferate so that a cluster or specific "clone" of cells was now in each well, generating antibodies, and secreting the highly specific antibodies from the hybridoma cells into the liquid growth media in those wells. Once the hybridoma cells had grown sufficiently and secreted enough antibodies to be tested, all that remained was to take samples from the growth media in each of the individual wells and determine whether the antibodies generated by the hybridoma cells in each well were capable of binding to the antigen of choice (Figure 14.3).

MONOCLONAL ANTIBODIES IN BIOMEDICAL RESEARCH

Antibodies, and particularly monoclonal antibodies with their inherent homogeneity, are a major engine that drives modern biomedical research. The global market size for research antibodies (monoclonal and polyclonal) has been estimated at well above 3 billion US dollars for 2019, with a significant share of that market coming from monoclonal antibodies. While monoclonal antibodies meet a vast array of research needs, among their key uses are: (1) the detection of protein–protein interactions through a variety of means, (2) the localization of proteins within cells and tissues, and (3) the detection and/or measurement of levels of proteins in the lysates of cells and tissues. These versatile functions of monoclonal antibodies in biomedical research enable an overwhelming percentage of the overall research done in laboratories across the globe. Indeed, the fact that these antibodies are "standardized" in that once a useful hybridoma has been identified, there is an endless supply of the *exact* same reagent, renders them an incredibly powerful tool for research. However, just as importantly, their use drives new discoveries that are then translated into new drugs and treatments. However, monoclonal antibodies are also *directly* used as treatments for a variety of diseases.

MONOCLONAL ANTIBODIES IN MEDICINE

Despite the emergence of monoclonal antibody technologies depicted by Kohler and Milstein in 1977 (Kohler and Milstein 1975) and the huge resulting medical promise, a major obstacle remained before this method could be advanced to clinical use; the monoclonal antibodies were derived primarily from mouse spleens or the spleens of other small mammals, such as hamsters and rats, and injecting or trying to use these antibodies for treatments in humans could evoke a massive immune response that would render them dangerous because the human immune system reacts to the presence of proteins from another species. Obviously, human spleens cannot be harvested to isolate human B cells for fusion with myeloma cells to generate human-derived hybridomas. However, the colossal significance of the discovery of monoclonal antibodies sparked researchers around the world to continue building on these findings. Just two years later, a new method was identified to generate *humanized* antibodies using a virus known as Epstein–Barr virus (EBV) to transform human B cells (Steinitz et al. 1977). These researchers demonstrated that once human B cells had been pre-selected to identify the ones that are relevant for interaction with a specific antigen, they could be transformed into an *immortal clone* (also called *immortalized*) that can continuously proliferate and endlessly generate the required monoclonal antibody. This was soon followed by discoveries that mouse antibodies could be *chimerized* or humanized by genetically fusing the Fab and hypervariable "Y-regions" regions of the mouse immunoglobulin protein that interact with antigens to the more constant human Fc "tail" of the antibody (see Figure 14.1), thus rendering them less likely to evoke an immune response when injected into the serum of humans (Morrison et al. 1984, Jones et al. 1986). These and other new technical advances continue to fuel the rapidly expanding clinical uses of monoclonal antibodies.

Monoclonal antibodies have a wide range of medical uses, primarily due to their remarkable specificity in binding to select antigen targets, thus reducing the level of toxicity and side effects that often occur for more conventional and less selective drugs that in addition to their specific targets can also inadvertently hit the wrong targets. As noted above, the versatility of monoclonal antibodies in the treatment of diseases is rapidly growing. One of the first successful monoclonal antibody treatments developed was for rheumatoid arthritis; in this illness, there is enhanced expression of a variety of pro-inflammatory proteins known as cytokines that contribute to the inflammation and progression of the disease. Since one such cytokine, known as Tumor Necrosis Factor α (TNFα), is a major regulator of many other inflammatory cytokines, the use of TNFα monoclonal antibodies became an important treatment for rheumatoid arthritis (Brennan et al. 1989, Feldmann and Maini 2003). Since then, TNFα monoclonal antibodies have also been used for a variety of other inflammatory diseases, including Crohn's disease and ulcerative colitis, psoriasis, and others (Shepard et al. 2017).

In addition to a variety of uses in combatting various inflammatory diseases, monoclonal antibodies have also been incredibly useful in treating many forms of cancer. Notably, humanized monoclonal antibodies (known as Trastuzumab or Herceptin) against the epidermal growth factor receptor known as HER2 have been effective in treating breast and other cancers. The mechanisms for Trastuzumab function are complex, but because breast cancer and certain other solid tumors frequently exhibit enhanced expression levels of the HER2 receptor, the monoclonal antibody is thought to function at several levels, inhibiting growth signals that are transmitted via the HER2 receptor into the cell, promoting an antibody-dependent immune response to kill tumor cells expressing high levels of HER2, and potentially increasing the sensitivity of the tumor cells to TNFα (Shepard et al. 2017).

Monoclonal antibodies are becoming treatments of choice for many blood cancers. One of the best known successes has been the use of an antibody against CD20, a protein found at unusually high levels on the surface of lymphocytes in a variety of lymphomas, a type of cancer that causes the massive proliferation of B or T lymphocytes (Chung 2019). The Fab regions of the anti-CD20 monoclonal antibodies (known as Rituximab, Ofatumumab, and a variety of others names) bind with high specificity to the CD20 on the surface of the B lymphoma cells, and couple these cells via their Fc regions to effector cells that carry out antibody-dependent cell cytotoxicity (ADCC), thus destroying the cancerous cells. In other cases where the type of lymphoma is a T cell lymphoma, such as in anaplastic large T cell lymphoma (ALCL), a surface protein known as CD30 is found at unusually high concentrations on the plasma membrane. In the case of T cells (unlike B cells), however, ADCC does not occur on its own when an antibody binds to the CD30 target found on these cells. Accordingly, scientists have coupled a toxin to the tail of the antibody, and once the monoclonal binds to the CD30 on the surface of these ALCL cells, it is internalized by receptor-mediated endocytosis (see Chapter 22) together with its conjugated toxin. The toxin currently used is a microtubule depolymerizing agent, which is delivered into the cancer cell and which it subsequently destroys with exquisite selectivity (van der Weyden et al. 2017). Other monoclonal antibody-based treatments in use include treatments for multiple myeloma, various leukemias, melanoma, head and neck squamous carcinoma, and many other cancers, in addition to treatments for asthma, Crohn's disease and ulcerative colitis, paroxysmal nocturnal hemoglobinuria, and other illnesses (Shepard et al. 2017).

Aside from their rapidly growing use for the treatment of diseases, just as importantly monoclonal antibodies have become an essential part of the physician's diagnostic toolkit. Indeed, monoclonal antibodies are the gold standard for assays to diagnose many types of infectious diseases, as well as cancers and other hormonal imbalances (Siddiqui 2010). The immense improvement over polyclonal antibodies comes from the reproducibility of each test, since the quality and affinity of the monoclonal antibodies are always the same, and because once a hybridoma clone has been generated, a renewable source of more of these exact same antibodies is always available. Compared to polyclonal antibodies, where a new rabbit and different reaction can lead to a variety of antibodies with varied specificity each time, monoclonal antibodies have completely revolutionized the way diseases are diagnosed. Even for entirely new diseases that surface, such as COVID-19, monoclonal antibodies have already been generated and are already in use to determine whether someone has been exposed to the virus and has developed an immune response to it. Thus, from a medical and public health standpoint, the uses of monoclonal antibodies as diagnostic tools are probably no less important than their role in the treatment of diseases. In any case, these two sides of the same "antibody-coin" further highlight the major significance of the discovery of antibodies, and the importance of basic science in their development.

15 Onto Oncogenes
*Discovering the Molecular Basis of Cancer****

In 1989, the Nobel Prize for Physiology or Medicine was awarded to J. Michael Bishop and Harold Varmus for their research on "oncogenes," a term coined by Robert Huebner that found its way into the scientific literature as early as 1958 (Rowe et al. 1958). Aptly, the Nobel description of the discoveries of Bishop and Varmus states that the prize was awarded for their "discovery of the cellular origin of retroviral oncogenes." Indeed, this is clearly distinct from simply "the discovery of oncogenes," and like many of the scientific discoveries detailed in this book, this one too was years and decades in the making. Moreover, consistent with other major biomedical findings made since the 1970s, the discovery of oncogenes and their role in tumorigenesis relied heavily on the great advances that were made in understanding molecular biology and genetic engineering.

Over 100 years ago in 1911, a scientist at the Rockefeller Institute in New York named Peyton Rous was the first researcher to discover and publish his findings demonstrating that viruses have the potential to cause cancer in infected hosts. His findings were predated by the study of Domenico Rigoni-Stern, who noticed that nuns in Verona, Italy, were far less susceptible to cervical cancer than married women (Scotto and Bailar 1969), leading him to the notion that the disease might be caused by a virus transmitted through sexual contact. However, while many school children today are vaccinated with a relatively new vaccine against the papilloma virus responsible for cervical cancer, this virus was not identified as being causative for the disease until 1983 (Durst et al. 1983). In contrast, the clear-cut connection demonstrated by Rous between viral infection and cancer was a turning point that eventually led to dramatic advances in understanding the molecular mechanisms of cancer.

Rous' work on avian viruses advanced rapidly when he was provided with a Plymouth Rock hen (one of the oldest breeds of American chickens) that bore a large spindle-cell sarcoma. Rous used cell-free tumor extracts from the hen and found that when he injected them into other fowl, this led to the induction of solid sarcoma tumors, and thus concluded that a "filterable agent" or virus was capable of transforming normal chicken

DOI: 10.1201/9781003202974-15

cells into malignant ones in healthy chickens (Rous 1911). Moreover, based on the ability of filterable agents from specific types of tumors to induce only that select tumor-type in healthy animals after injection (Rous and Murphy 1914), Rous essentially laid out the groundwork for the future argument that viruses transform based on their genetic material. Based on his novel findings, Rous did something fairly typical in the scientific world; he incorrectly inferred that the transformation induced by the virus (that would later become known as Rous sarcoma virus (RSV) in his honor) was representative of all cancers and he advocated that somatic mutations could not cause cancer (Rous 1959). Ironically, he did not know that the RSV is a very potent mutagen that can induce somatic mutations in cellular genes to transform cells. Despite his mistaken extrapolations—and it should be noted that to make breakthroughs, scientists should not be afraid of being wrong—his findings were nonetheless highly significant and led to the development of a budding new field known as *tumor virology*. But however important and illuminating Rous' research was, an understanding of the mechanisms that underlie the viral transformation of cells remained many decades away.

Harry Rubin, a veterinarian by training and virologist at Caltech in California, made important findings in the mid-1950s when he showed that all cells in an RSV tumor were able to release virus without necessarily causing cell death, leading Rubin to propose that the viral association with the cells may be an integral part of the oncogenic mechanism (Rubin 1955). This was followed up by additional important developments in discovering the molecular basis of oncogenes from the experiments done by Howard Temin (who later shared the Nobel Prize in 1975 with David Baltimore and Renato Dulbecco for co-discovering the reverse transcriptase enzyme that would revolutionize molecular biology) together with Harry Rubin at Caltech.

Temin and Rubin pioneered a quantitative and sorely needed cell culture-based assay that they termed the *focus assay* thus facilitating quantification of oncogenic activity, an assay that would become the gold standard for measuring and quantifying transformation activity (Temin and Rubin 1958). According to Peter Vogt (Vogt 2010), who worked in the Rubin lab and was an integral player in the key discoveries made in defining the RSV genes, their size, order, and the position of the crucial *src* gene ultimately responsible for transformation (Wang et al. 1975), there was not always ideal harmony between Temin and Rubin in the laboratory. Vogt describes Rubin as a cautious and highly rigorous scientist who shied away from making any speculations that veered beyond what could be concretely concluded from the data (Vogt 2010). Temin, on the other hand, possessed vision and great imagination. Intriguingly, this kind of scientific team is analogous in the detective world to pairings such as that of Sherlock Holmes and Dr. Watson or to the great Hercule Poirot and his sidekick Captain Hastings. Having one researcher who excels in a type of rigorous, direct, and focused research, while the other is continually keeping his eye on the ball for a grand-slam, a dogma-shattering new way of thinking, has often been beneficial to the scientific world in the long run. In this case, it was Temin's vision of the provirus (published together with Satoshi Mizutani) (Temin and Mizutani 1970), along with work done by David Baltimore at the Massachusetts Institute of Technology (Baltimore 1970) and the potential for RNA viruses to use DNA as an intermediary that ultimately led years later to the discovery of reverse transcriptase—and of course, the start of the era of genetic engineering.

Research on Rous sarcoma virus and the mechanism of virally transduced cancer was advancing on multiple fronts. Around the time that Temin was first publicizing his revolutionary game-changing hypothesis that RNA viruses might use DNA as an intermediary before ultimately leading to mRNA transcription and protein translation, Steve Martin at the University of California at Berkeley identified a temperature-sensitive mutant of RSV (Martin 1970). This means that at the non-permissive temperature, while his virus would replicate normally, it was incapable of transforming cells into continuously dividing cancer cells. From a practical standpoint, this temperature-sensitive RSV mutant virus now presented researchers with an outstanding tool to determine what differences exist between the transforming RSV and the non-transforming temperature-sensitive RSV mutant—in an attempt to identify the specific requirements of the virus for its transforming ability. Through a series of experiments comparing the two viruses, Peter Vogt and Peter Duesberg working at the University of California at Berkeley demonstrated that while both viruses shared the genes needed for replication (known as *gag*, *pol*, and *env*, which stand for "group of antigens," "polymerase," and "envelope protein," respectively), the temperature-sensitive RSV that was unable to transform was missing a segment of RNA at the 3' end of the virus (Duesberg and Vogt 1970). This missing region, ultimately dubbed *src*, codes for the viral protein required for the transformation of cells.

Despite the incredible earth-shattering findings initiated by Temin and others that resulted in the smashing of the central dogma of biology, showing that RNA from RNA viruses is indeed reverse transcribed back into DNA before its transcription to mRNA and translation into protein, the scientific world probably could not have predicted that within the span of a few short years a second earth-shattering discovery would also come from the realm of virology. In a 1976 paper published in the journal *Nature*, Dominique Stehelin, Harold Varmus, and J. Michael Bishop of the

University of California at San Francisco, and Peter Vogt the University of California at Los Angeles discovered that the *src* gene of RSV (now known as *v-src* for viral *src*) is a transduced allele of a gene that exists within normal host cells (*c-src* for cellular *src*) (Stehelin et al. 1976). This turned the fields of virology and cancer research on their heads, because it meant that the potential for cancer transformation is found in every cell, and that the *RSV likely initially acquired the gene from host cells* through some form of genetic recombination. Indeed, the realization that these *proto-oncogenes* are found in human cells (and not just viruses) and that simple mutations or any form of dysregulation of the normal *c-src* in the host genome might turn them into full-blown cancer-causing oncogenes (even without any viral involvement) has led to a major shift in biomedical and scientific thinking.

As scientists began to understand the relationship between genes and proteins, to truly comprehend oncogene function it became imperative to identify the protein that is coded by the *src* gene and to decipher its function and the mechanism by which it confers cellular transformation. Initial attempts included cell-free *in vitro* translation assays, in which the viral RNA would be provided with the machinery for its translation to protein, but most of these attempts led to identification of only segments of the protein coded for by the *src* gene (Beemon and Hunter 1977, Kamine and Buchanan 1977). Ultimately, identification of the protein product of the *src* gene came from a completely different approach—an immunological approach rather than a genetic one. The strategy was to inject rabbits and induce tumors with a form of RSV, on the assumption that the rabbits would develop an immune response to RSV proteins, including the unknown *src* gene product. Eventually, using lysates extracted from RSV-transformed cells, a specific 60 kilodalton protein attached to phosphate groups (a phosphoprotein) was identified as the protein coded for by the *src* gene and it was termed Src (Brugge and Erikson 1977). Ultimately, studies by the laboratories of Tony Hunter, J. Michael Bishop, and others led to the discovery that the v-Src protein is a kinase that phosphorylates proteins on tyrosine residues—a discovery with immense implications for the entire field of researchers working on *signal transduction* in addition to those studying Src and the viral transformation of cells (reviewed in (Martin 2004)).

While *src* is an incredibly important oncogene, and the first one discovered, it certainly is not the only oncogene (although the focus of this chapter will be limited to a select few). In the early 1960s, Jennifer Harvey at the Cancer Research Department of the London Hospital Research Laboratories was working on a virus known as Moloney's leukemogenic virus (MLV) that causes leukemia in mice and rats. Astutely, she noticed that when she inoculated animals with plasma from infected animals that contained the leukemia virus, in addition to developing the murine leukemia as expected, some of the mice also developed solid tumors (Harvey 1964). Almost in parallel, Werner Kirsten, a pathologist and researcher at the University of Chicago, discovered a related virus with similar transforming ability (Kirsten and Mayer 1967). These murine viruses later became known as the H-*ras* (Harvey Ras) and K-*ras* (Kirsten Ras) murine leukemia viruses, respectively.

In 1974, Edward Scolnick and his co-workers at the National Institutes of Health in Bethesda, Maryland, predicted that the transformative ability of the H-*ras* and K-*ras* viruses might result from the virus being able to co-opt a host gene into its own viral genome, where the gene is mutated into an oncogenic form (Scolnick and Parks 1974, Cox and Der 2010). The formal proof of this idea, that viral oncogenes represent host genes that have been plucked by the virus and have undergone mutation into an oncogenic form, would await the aforementioned discoveries by Bishop and his colleagues, followed shortly thereafter by additional studies where Scolnick and his co-workers sequenced the *ras* genes and showed that they indeed originated from normal vertebrate genes (Ellis et al. 1981). In the meantime, Scolnick and his colleagues did provide the first data supporting the idea that Kirsten sarcoma virus arose through a process of recombination between Kirsten murine leukemia virus and nucleic acid sequences found in rat cells (Scolnick et al. 1973). Interestingly, as the mammalian *RAS* genes were beginning to be dissected by researchers, they were initially considered to be variants of the *SRC* gene, and were sometimes referred to as the p21 Src due to their apparent molecular weight of 21 kilodalton as opposed to 60 kilodalton for pSrc (Shih, Weeks et al. 1979).

One of the most significant advances in the field of oncogene research—indeed in any area of biomedical research—was a technological one that even today continues to drive biomedical research forward. A major limitation in the research carried out in the early 1970s, despite advances in molecular biology and the start of the era of genetic engineering, was the inability to easily assess gene function in cells. A number of laboratories during this time began toying with the idea of being able to express foreign DNA in cultured cells to determine how the expression of the protein coded for by the introduced DNA affected cell function (Weiss 2020, Malumbres and Barbacid 2003). A number of studies were beginning to develop new methods to introduce foreign DNA into cells (Hill and Hillova 1972, Miller et al. 1979, Wigler et al. 1978), which ultimately led to a breakthrough by Robert Weinberg and his co-workers by providing a new platform to determine if specific genes were oncogenic and caused cellular transformation. In these studies, the Weinberg group used a well-characterized cell line known as NIH3T3 cells and determined whether newly introduced DNA into these cells caused their transformation to continuously dividing cancer-like cells. They isolated genomic DNA from chemically transformed cells and showed that when this DNA from the transformed cells was introduced into the NIH3T3 cells, it was capable of transforming them (Shih, Shilo et al. 1979). Additional studies further supported the ability of foreign DNA to transform cells (Perucho et al. 1981), and ultimately a transforming gene was cloned from bladder carcinoma cells (Pulciani et al. 1982, Goldfarb et al. 1982, Shih and Weinberg 1982). Remarkably, it was not until the early 1980s that a combination of studies from several laboratories finally unambiguously demonstrated that the specific DNA isolated from NIH3T3 cells that had been transformed with DNA from tumor cells showed a match for the viral H-*ras* and K-*ras* oncogenes (Der, Krontiris, and Cooper 1982, Parada et al. 1982, Santos et al. 1982). Shortly afterward, a third transforming *ras* gene, N-*ras* was added to the list of viral oncogenes with human counterparts (Parada and Weinberg 1983, Shimizu et al. 1983).

Over the next decade, important discoveries were made in understanding the cellular functions of the Ras proteins. Unlike the Src protein, which is a kinase responsible for phosphorylation (in other words, the kinase chemically modifies select amino acids of a protein by the covalent addition of a phosphate group to them), the smaller Ras proteins were not kinases, but turned out to bind to guanine diphosphate (GDP) and guanine triphosphate (GTP) (Scolnick, Papageorge, and Shih 1979). Since that time, many proteins have been discovered with similar three-dimensional structures to Ras as well as homologous sequences, designating an entire family of Ras-like proteins (Cox and Der 2010). Like Ras, most of these Ras-like proteins cycle between GDP and GTP binding, and typically associate with membranes when they are in their GTP-bound state, but not when they are GDP-bound. Indeed, studies also found that it is the GTP-bound form of Ras that primarily serves as an oncogenic protein (Gibbs et al. 1984), and mutations in Ras that prevent the hydrolysis of GTP to GDP are frequent in various cancers (Brown et al. 1984, Sistonen and Alitalo 1986). It was also found that Ras proteins are linked to the inner leaflet of the plasma membrane, suggesting that they function in delivering signals from outside the cell to the cell interior (Hurley et al. 1984). Since that time, numerous studies have addressed the complex myriad of signaling pathways in which Ras proteins transduce signals from the cell exterior, as well as from endosomal membranes within the cell, ultimately leading to gene expression and cell proliferation (Malumbres and Barbacid 2003).

As with most basic biomedical discoveries, the progression from abstract knowledge and an understanding of oncogene function to the translation of that knowledge into actual clinical advances is a slow but steady process. Over the years it became clear that there is a family of almost a dozen Src-related protein kinases (including Blk, Fgr, Fyn, Hck, Lck, Lyn, Src, Yes, and Yrk) that shares considerable sequence and functional homology (Manning et al. 2002), and whose expression can either be ubiquitous or restricted to specific tissues and cells (Parsons and Parsons 2004). Through the work of numerous laboratories, using biochemical, structural molecular, and cell biological methods, many Src-family inhibitors have been identified and developed over the years. At the start of 2021, there were over 50 clinical trials listed at the National Institutes of Health's ClinicalTrials.gov website that have either used or are in the process of using a Src inhibitor to assess the treatment of various forms of cancer. For example, the "second generation" inhibitor known as dasatinib impedes Src function, is approved for use in patients diagnosed with chronic myeloid leukemia (Rivera-Torres and San Jose 2019), and is currently being tested in clinical trials for a variety of additional cancers including ones that are typified by solid tumors. Another Src inhibitor known as saracatinib is under clinical trials for forms of lung cancer as well as potential use in Alzheimer's disease and other neurodegenerative disorders. The search for potent and highly specific inhibitors of Src and other tyrosine kinases is a very active area of cancer research and will likely continue to yield new tools in the fight against cancer as well as other diseases.

In contrast to the success that work with Src inhibitors has yielded, Ras inhibition has remained an elusive target and for years Ras has been considered an "undruggable target," owing in part to the lack of sufficiently deep pockets in the three-dimensional structure of the protein that would facilitate the binding of potential inhibitors (O'Bryan 2019). However, as Ras mutations have been observed in ~30% of human cancers (O'Bryan 2019), it remains a crucial target for researchers to nonetheless try to address. While initial attempts were designed, mostly unsuccessfully, to block Ras function directly, more recently researchers have turned their attention to potential opportunities to block Ras localization to the plasma membrane where it needs to reside in order to function and transmit signals, or to "effectors" of Ras that are activated by Ras to propagate signals onward (or "downstream") within the cell. A recent study suggests that the oncogenic KRas mutant (with the glycine residue at position number 12 substituted with a cysteine) may be "druggable" with a compound known ARS-1620 (Janes et al. 2018), but it may be many years until Ras inhibitors are included in the repertoire of anti-cancer drugs used in clinical settings.

A SUPPRESSOR SURPRISE—TUMOR SUPPRESSORS TO THE RESCUE

As the work on oncogenes was progressing, in 1979 three papers were published that addressed the oncogenic potential of Simian 40 (SV40) and polyoma viruses, each of which supported the notion that a 53 kilodalton protein that had previously been considered to be a part of the SV40 virus was actually a cellular protein (Kress et al. 1979, Linzer and Levine 1979, Linzer, Maltzman, and Levine 1979, Lane and Crawford 1979). Almost in parallel, immunology researchers with a different approach found that SV40-induced tumors led to an antibody-based immune response to a 53 kilodalton protein (DeLeo et al. 1979). This 53 kilodalton protein that was suspected at the time of harboring oncogenic activity turned out to be the very same cellular protein which was ultimately termed p53 (Crawford 1983).

The discovery of p53 and its initial categorization as an oncogene before its dramatic reclassification as a *tumor suppressor gene* highlights what researcher Thierry Soussi describes as the drawback of scientific paradigms (Soussi 2010). In laying out the initial studies after the discovery of p53, Thierry demonstrates how easily p53 was ascribed oncogenic functions, a situation that became further complicated with each new paper that was published—until finally the contradictions in experimental results could no longer fit the oncogene dogma. Thierry notes that between 1985 and 1988 there were 131 papers that were published on p53 that could not support p53 function as an oncogene, and therefore interest in p53 research began to wane (Soussi 2010).

In 1989, research led by Arnie Levine and his laboratory at Princeton, New Jersey, was responsible for cloning the normal mouse p53 gene. Their research showed that it did not behave like an oncogene, but rather it *suppressed* cellular transformation, contradicting the many published papers maintaining that it acts as an oncogene (Finlay, Hinds, and Levine 1989). Coupled with studies showing that mutations in the cellular p53 gene led to various cancers, including colorectal cancer (Baker et al. 1989) and lung cancer (Takahashi et al. 1989), the work from the Levine laboratory finally spurred a dogmatic shift in thinking about p53 function and cancer researchers revised their previous assessments of p53 as an oncogene and began to call it a tumor suppressor gene.

The major advances in technology since the dawn of the twenty-first century that led to the sequencing of the entire human genome have also facilitated large-scale sequencing to identify mutations in genes that are present in a variety of diseases, especially in cancers. Based on such studies, it was shown that the p53 gene is mutated in many forms of cancer, including 96% of ovarian serous carcinomas (Cancer Genome Atlas Research 2011), 75% of pancreatic cancers (Yachida et al. 2012), and 54% of invasive breast carcinoma cancers (Shah et al. 2012). Given the preponderance of evidence in support of the notion that loss of p53 (or loss of its normal function) is key to many forms of cancer, over time this basic finding has led to attempts to capitalize on this understanding and "rescue" the impaired p53 function with therapy. One example of such a therapeutic strategy derived from knowledge that normal p53 function prevents tumorigenesis is the development of a small molecule known as PRIMA-1 or PRIMA-1MET/APR246, which appears to facilitate a return of p53 to its active conformation (Bykov et al. 2002). Resulting clinical studies in humans with these drugs show promise for a variety of different p53-related cancers, but the overall efficacy is still being evaluated in these trials (Blandino and Di Agostino 2018).

Oncogenes and their counterpart tumor suppressor genes are additional examples of how basic scientific understanding takes the concentrated effort of many laboratories. Importantly, it is often necessary to overturn inbred dogma in order to make significant advances. In the case of oncogenes, the discovery that viruses encode genes that were originally captured from host cells was an illuminating example. Alternatively, in the case of tumor suppressor genes, a second leap had to be made to propose that the normal p53 gene product actually provided protection against tumorigenesis. In both cases—for classic oncogenes and tumor suppressor genes—basic knowledge preceded clinical trials by 25–30 years. So while Martin Luther King, Jr., said in his speech titled "Remaining Awake Through a Great Revolution" that "The arc of the moral universe is long, but it bends toward justice" (Martin Luther King 1968), one might modify that epic statement in the case of biomedical science and proclaim: "The arc of basic biomedical discovery is long, but it bends toward clinical impact."

16 The Age of Angiogenesis

*Discovering How Blood Vessels Are Generated**

Many of the great biomedical discoveries described in this book relate to the finding or identification of a single important molecule. For example, the discovery of insulin (Chapter 12), the cause of sickle cell anemia (Chapter 9), or the characterization of antibodies (Chapter 14). Others relate to the discovery of key organelles within the cell, such as the Golgi complex (Chapter 19), the lysosome (Chapter 20), and the primary cilium (Chapter 18). A number of discoveries described within are concerned with pathways, such as receptor-mediated internalization (Chapter 22) or proteasomal degradation (Chapter 21). However, a unique category is the discovery of a physiological process that occurs within an organism—especially a highly significant one whose initiation can ultimately be traced back and attributed to key individual molecules. Such is the case in the discovery of angiogenesis.

Angiogenesis is the physiological process in which new blood vessels are generated in the body, and it is a highly regulated process that typically occurs in adults only in the course of tissue repair/growth, and for women, in the reproductive cycle. Accordingly, dysregulation of angiogenesis is a hallmark of various diseases, particularly in the case of solid tumors, although the significance of angiogenesis in cancer has only been appreciated starting from the 1970s.

An investigation of the origin of the term *angiogenesis* reveals that it was likely first used by the famous Scottish anatomist, surgeon, and scientist, John Hunter, in the 1700s (Lenzi, Bocci, and Natale 2016). Hunter was a fascinating and extremely versatile scientist who collaborated with Edward Jenner, the latter being credited with developing the first vaccine against smallpox (see Chapter 14). Hunter reportedly used the term angiogenesis to describe the blood vessels developing in reindeer antlers in 1787 (Lenzi, Bocci, and Natale 2016). However, Hunter was also known for illegally snatching the body of Charles Byrnes, a giant of approximately

DOI: 10.1201/9781003202974-16

7-foot 7-inch stature, for display in his own private museum collection. Hunter apparently offered Byrnes and his family money for his skeleton after his imminent death, but having been refused, Hunter proceeded to steal the giant's body before Byrne's family and friends could bury the giant at sea (Doyal and Muinzer 2011).

It was nearly a hundred years until the next angiogenesis pioneers made significant new observations and discoveries. Karl Thiersch, working in Leipzig, Germany, showed that new blood vessels are generated from existing capillaries, and his descriptions of angiogenesis laid the foundation for future notions that blood vessels that are formed in an irregular manner might be consistent with tumorigenesis (Thiersch 1865). Almost at the same time, Rudolph Virchow in Berlin noticed that blood vessels within tumors are essential for their blood supply (Virchow 1863). Additionally, it was observed that newer areas of tumors commonly had enhanced networks of vasculature, whereas older areas of the tumor were prone to having atrophied blood vessels (Billroth 1871)—providing more circumstantial evidence that angiogenesis is required for tumorigenesis.

In the early 1900s, researchers were beginning to connect the dots between tumor growth and the expansion of blood vessels. Studies from the laboratory of Edwin Goldmann, a South Africa–born researcher who trained under Paul Ehrlich in Berlin and was also the discoverer of the blood–brain barrier, found that tumors disrupted the normal pattern of blood vessels and that massive angiogenesis occurs during tumor formation (Goldmann 1908). Interestingly, Warren Lewis, an extremely versatile researcher who was among the first to image cultured cells by video microscopy (Lewis 1936) (see Chapter 22), used the experimental systems from these studies to derive an incorrect conclusion—that blood vessels do not determine the growth of the tumor, but rather that the tumor affects the growth of the blood vessels. While hypotheses and models need to be proposed and interpreted for science to advance—even when they are wrong—in this case Warren's interpretations may have made it significantly harder for Judah Folkman, decades later, to convince his peers that blood vessels are required for tumor growth.

In the mid-1900s, an American pathologist named Harry Greene built on previous work and began to study transplantations. In the course of his research he noticed an important relationship between tumors and vascularization. His main observation was that when tumors were transplanted into the eyes of animals, the tumors would grow rapidly when neovascularization occurred, but in cases when the tumor did not grow, neovascularization seldom occurred (Greene 1941). However, it seems that Greene's work on tumors and transplantation was not particularly well known at the time, and he was better known for his insistence that smoking did not cause lung cancer. He died from lung cancer some years later in 1969, having been a heavy smoker (Stephenson 2013).

It was not until the early 1960s when Moses Judah Folkman, a surgical resident who was drafted into the navy and was working at the Naval Medical Research Institute in Bethesda, Maryland, again made the connection that tumor growth seemed to be directly connected to tiny new networks of blood vessels in the proximity of the tumor (Folkman, Long, and Becker 1962). Over time, based on his clinical experiences with retinoblastomas, Folkman developed his theory of angiogenesis in which he stipulated that tumor growth is absolutely dependent on angiogenesis, and that beyond a distance of 2 mm, the diffusion of oxygen and nutrients to supply a growing tumor would be insufficient without the generation of new blood vessels (Folkman 1971). With this new hypothesis, Folkman expanded his ideas further and proposed that without angiogenesis, tumors would become restricted and not continue to grow, that if angiogenesis were to be prevented, tumors would stop growing. Finally, Folkman proposed that it was likely that tumors secrete soluble factors to promote angiogenesis in a type of "positive feedback loop" to promote tumor growth.

The existence of an actual angiogenic factor—a molecular substance secreted by tumor cells that would induce the generation of new blood vessels—was the key to definitively proving Folkman's hypotheses. The main problem that had to be overcome to make this possible was the development of an *in vitro* system of endothelial cells. The reason an *in vitro* cell system was necessary is that once researchers had performed biochemical separations or purifications of molecules from tumor cells, they needed a fairly simple and rapid system in which the isolated molecules could be tested to determine if they are angiogenic in nature. Obviously, the older system of injecting potential angiogenic factors into animals and waiting weeks or months to assess the outcome was simply not feasible. The problem was that endothelial cells (the cells that comprise blood vessels) were notoriously hard to passage in culture. These difficulties were temporarily sidestepped by Folkman and his colleagues when they developed a robust method for implanting select purified proteins into the cornea of mice within polymer pellets to assess whether neovascularization occurred (Muthukkaruppan and Auerbach 1979, Langer and Folkman 1976). Soon after, endothelial cell culture was standardized and Folkman's laboratory and others began to use *in vitro* systems to assess whether specific proteins were indeed angiogenesis factors, as an initial rapid method before proving the substances were effective in animal models.

It was not until the 1980s that evidence from multiple laboratories, including Folkman's own laboratory, began to mount supporting the existence of

Folkman's angiogenic factors. The first one, known as vascular endothelial growth factor (VEGF), also termed "vascular permeability factor" (VPF) as well as "vasculotropin," was identified in parallel by several laboratories (Ferrara and Henzel 1989, Keck et al. 1989, Rosenthal et al. 1990). Other angiogenesis factors have also been discovered, including members of the fibroblast growth factor (FGF) family (Cross and Claesson-Welsh 2001) and the angiopoietin family (Ribatti, Vacca, and Presta 2000), although of these angiogenesis factors VEGF remains one of the prime targets for clinical use.

In the early 1990s, several groups began to identify the receptors to which VEGF and angiogenic factors bind, with the first two receptors identified known as VEGF receptor-1 (VEGFR-1) and VEGF receptor-2 (VEGFR-2) (Terman et al. 1992, de Vries et al. 1992, Shibuya et al. 1990). While VEGFR-2 appears to be the main VEGF signaling receptor, all three VEGF receptors are tyrosine kinases with similar structures and partially overlapping functions. The discovery of these VEGF receptors provided an additional avenue for researchers to try and interfere with the process of angiogenesis and thus limit the spread of tumors.

Dozens of angiogenesis inhibitors have been identified since the discovery of VEGF and the initial angiogenesis factors, and they can be categorized either as "direct" or "indirect" inhibitors, depending on whether they interfere with the angiogenesis factors themselves or with their target receptors and affiliated downstream events. Direct inhibitors include angiostatin, avastin, endostatin, and thrombospondin, whereas indirect inhibitors target the VEGF receptors as well as targeting immunomodulatory proteins such as interferon-alpha (Ribatti 2014). In the decade from 1990 to 2000, infant hemangioma patients began to receive treatment with interferon-alpha, and the vast majority of these infants responded successfully while mortality dropped by ten-fold (Stephenson 2013). However, despite these obvious successes in harnessing basic science and capitalizing on inhibitors of angiogenic factors to target select cancers, other malignancies remain much harder to address. This may be in part because many tumors tend to "co-opt" existing blood vessels for their own endeavors, rather than generating completely new ones (Kuczynski and Reynolds 2020, Kuczynski et al. 2019). Nonetheless, there are many dramatic successes derived from inhibiting angiogenesis in a variety of cancers beyond pediatric hemangiomas.

Today drugs that prevent or slow angiogenesis are among the most common type of treatments for multiple types of cancer, and many of them are designed to block the effects of VEGF (Meadows and Hurwitz 2012). The first VEGF inhibitor approved for cancer treatment is known as

Bevacizumab—a monoclonal antibody that specifically binds and neutralizes circulating VEGF. This "humanized" antibody (engineered to avoid an immune reaction against antibodies from different species) is currently being used for the treatment of several types of cancer, including ovarian, gastric, prostate, urinary, lymphomas, glioblastomas, and others, often in combination with other drugs (Meadows and Hurwitz 2012). A variety of selective tyrosine kinase inhibitors that inhibit VEGF receptors have also either been recently approved or are in the process of development by different drug companies, and clinical trials are well underway to address their efficacy (Meadows and Hurwitz 2012). Accordingly, the time frame from the basic discovery of angiogenesis factors in the early 1980s to approval of drugs that take advantage of this knowledge and target the angiogenesis factors is about 30 years, highlighting the significance of making new basic discoveries now for improved clinical outcomes in 2050 and beyond.

In summary, despite Folkman's work and growing evidence that vascularization is required for continued tumor growth, as well as the inherent common sense in these ideas, acceptance of the notion that angiogenesis plays a key role in tumor growth was very slow to become established. While in retrospect some new scientific ideas often appear brilliantly clear, the history of science is rife with cases where new dogma has a hard time taking root. The reasons for this are complex and likely include a combination of the following rationales: (1) some scientists do not like to abandon long-held theories, because they (wrongly) think this makes them appear to be poor researchers for "barking up the wrong tree," (2) it is easier, simpler, and requires less effort to maintain a previously made hypothesis than to propose or even support a new one, (3) scientists are by nature conservative and skeptics, which is an important feature of scientific training, (4) the nature of establishing new dogma is that novel ideas come from new evidence, and individual pieces of such evidence often filter in a little bit at a time, so that it often takes many years until a critical mass of new supporting evidence is reached. In addition to these issues, one must also take into consideration that at least in the early 1970s (compared to 50 years later) scientific communication was not instantaneous by any means. The publication and dissemination of scientific papers took months even after their acceptance in the journals, and the scientific literature was not readily accessible on laptops, which were of course nonexistent at the time. This is unlike 2021, where exciting new discoveries are hyped by the journals immediately upon publication, instantaneously deposited directly in scientists' email boxes, and even promoted on social media. Accordingly, the dissemination of knowledge even in the 1970s was a considerably slower process.

17 Telling the Tale of the Telomere and Telomerase

*The End of the End-Replication Problem****

It is possible that the history of the discovery of telomeres, along with telomerase, the enzyme that generates and maintains these important regions at the ends of chromosomes, began with studies by Hermann Muller and Barbara McClintock, who won the Nobel Prizes in 1945 and 1983, respectively, for their studies on the effects of X-ray radiation on mutagenesis and for the discovery of mobile genetic elements. Muller and McClintock each proposed the existence of a structure at the end of chromosomes that would serve to protect them and ensure accurate segregation upon cell division (McClintock 1939, Muller 1938). However, it was not until the 1970s that Alexei Olovnikov realized that telomeric DNA at the ends of the chromosomes could potentially explain an interesting and important phenomenon known as the *Hayflick limit*: that human cells in culture can only undergo a certain number of replications (Hayflick and Moorhead 1961). Indeed, Olovnikov made the connection between the serial shortening of the chromosomes (particularly in their telomeric regions) with each cell division and their reaching the Hayflick limit and being unable to divide further (Olovnikov 1973).

The telomeres protect the chromosomes from a variety of detrimental events, including rearrangement, fusing of the chromosome ends, DNA damage, and recombination, and are typically comprised of several thousand sequence repeats of the nucleotides TTAGGG (Relitti et al. 2020). Upon each cycle of cell division, 50–200 of these nucleotides are lost at the telomere, and it is the cumulative loss of these nucleotides that is thought to lead to the enactment of the Hayflick limit and ultimately induce the prevention of further cell division. This inability of the DNA polymerases to copy the very ends of the chromosomes (where the telomeres reside) due to the directionality of DNA replication led Olovnikov and later James

DOI: 10.1201/9781003202974-17

Watson to their framing of what became known as "The End-Replication Problem" (Olovnikov 1973, Watson 1972). Accordingly, it was hypothesized that there must be an existing biological mechanism to prevent the continued shortening of telomeres—otherwise organisms would not be able to survive and reproduce.

In 2009, the Nobel Prize was awarded to Elizabeth Blackburn, her student Carol Greider, and her collaborator Jack Szostak for their discovery of "how chromosomes are protected by telomeres and the enzyme telomerase." In a classic series of experiments, Blackburn identified the telomere sequences in the model organism *Tetrahymena*, and she and Szostak then coupled the *Tetrahymena* telomere sequences to a linear chromosome-like piece of DNA ("mini-chromosome") that Szostak had previously observed to undergo rapid degradation in the yeast he was working on. Remarkably, when coupled with Blackburn's telomere DNA, Szostak's "mini-chromosome" became protected from degradation in the yeast cells (Szostak and Blackburn 1982).

Despite these intriguing findings, several puzzling questions remained unanswered. First, how are telomeres maintained over time when they appear to be progressively shortened with each cell division cycle? Moreover, if the telomeres are progressively shortened, then how do cells continually divide and give rise to daughter cells and offspring organisms? Blackburn and Greider hypothesized that perhaps a "special" polymerase exists that is distinct from the DNA polymerase that normally replicates DNA during cell division. This idea turned out to be correct and was validated, and the duo succeeded in purifying and identifying *Telomerase*; a complex that contains both RNA and proteins and is capable of extending the telomeric sequences (Greider and Blackburn 1985, 1989). Indeed, the RNA portion of telomerase turned out to be complementary to the telomeric repeat sequences, and thus capable of serving as a template for the enzyme to copy this sequence. Shortly afterward, telomeres were identified in human cells (Morin 1989), and it was proposed that cells that are immortalized might have more telomerase to allow them to grow indefinitely in culture. Subsequently, it was found that ovarian cancer cells contained telomerase activity, whereas normal cells did not, lending support to the idea that excess telomerase function might lead to oncogenesis (Counter et al. 1994). Consistent with these findings, mouse models engineered to have enhanced telomerase activity had higher rates of tumor generation and mortality (Gonzalez-Suarez et al. 2005). In addition, given that short telomeres can interfere with cell division, one might anticipate that cancer cells have enhanced telomerase activity. This turned out to be the case, and studies demonstrated that most cancer cells typically activate telomerase activity during tumorigenesis, thus ensuring that telomere length does not dip below a minimal level so that continued cell division is possible (Counter et al. 1994, Kim et al. 1994).

Given the close relationship between telomere length, telomerase, and cell division, it is not surprising that these important basic findings that enhance understanding of the fundamental mechanisms of cell division have been rapidly harnessed for therapeutics. Since telomerase activity is extremely low in normal adult tissues, and high in most cancer cells, telomerase activity is being used as a cancer biomarker for both disease and cancer prognosis (Xu and Goldkorn 2016). A key technological advance that stemmed from the discovery of PCR (see Chapter 6) was a method called the *telomeric repeat amplification protocol (TRAP)* in which a PCR-based amplification could be used to measure levels of telomerase activity in a cell lysate (Kim et al. 1994). Both detection of the existence of such telomerase activity with TRAP and the measurement of mRNA for the telomerase enzyme TERT are now being used as significant biomarkers for a range of different cancers (Xu and Goldkorn 2016). This serves as another example of the relationship between technology (or methods) development and the translation of basic research into clinical advances.

As of 2020, a number of clinical trials were underway to test the usefulness of inhibitors that block or decrease telomerase activity. Imetelstat is one such inhibitor that binds directly and with high affinity to the RNA fraction of telomerase in a very specific manner, and thus inhibits its function. Importantly, studies have demonstrated that Imetelstat is effective in inhibiting telomerase activity in a variety of tumors, including breast cancer, liver cancer, brain cancer, blood cancers, hematological cancers, and others, and it has been evaluated in at least ten different clinical trials thus far, alone or in combination with other drugs, for a number of different malignancies (Relitti et al. 2020). An additional telomerase inhibitor known as BIBR1532 that blocks the function of the human TERT (hTERT) enzyme also shows promise but is still awaiting clinical trials.

An interesting new possibility for therapeutics is one that combines knowledge from the basic science of telomeres and telomerase, cell biology, and protein degradation, and couples it with immunology and the science of vaccines. This stems from basic observations that in cells expressing hTERT (that have telomerase activity), there is some degradation of the telomerase enzyme and "presentation" of these degraded hTERT peptides by major histocompatibility complex molecules on the cell surface to cytotoxic T cells of the immune system for recognition and targeting (Minev et al. 2000). Accordingly, these studies led to the development of Phase 1 vaccine trials to immunize patients against hTERT. The goal of this study was to activate the immune system so that it will seek out and destroy cells that present hTERT peptide, with the understanding that these are primarily tumor cells.

A number of trials have been done with a variety of different vaccine systems, and one of the most promising is by inoculating with a 16-residue hTERT sequence from the active site of the enzyme. This sequence, known as hTERT 611-626 (GV1001), is in Phase 1–3 trials for different cancers, including non-small cell lung cancer, pancreatic cancer, and malignant melanoma.

While perhaps the basic findings and lessons learned from telomere biology and telomerase have yet to realize their full clinical potential, the biology of telomeres is another excellent example of how a fundamental understanding of the key biological and biochemical events within the cell eventually leads to the development of drugs and cures. Consistent with many of the other great biomedical discoveries described, the timeline for developing knowledge of telomere biology into clinical use has been decades long and it has required the perseverance of numerous laboratories, adding both small and large pieces to the puzzle. No less important is the concurrent development of technology, as the use of the TRAP PCR-based amplification system to track telomerase activity contributed significantly to the translation of basic telomere science into pharmaceutical enterprise.

18 The Primary Cilium *Novel Functions for an Old Organelle**

The primary cilium is a fascinating and breathtakingly beautiful organelle (Figure 18.1) whose true function was not discovered until the past 20 years. Indeed, the primary cilium serves as an excellent reminder of how great discoveries in basic science—including those whose significance takes decades, if not centuries, to unravel—can be incredibly significant to the clinical world. There are dozens of diseases known as *ciliopathies* that result from aberrant proteins that are involved in a variety of functions related to the primary cilium. For example, mutations in genes that code for proteins involved in intraflagellar transport or the movement of materials along the cilia often cause heterotaxia (misaligned heart symmetries) or polycystic kidney disease. The ciliopathies that result from impaired primary ciliogenesis include Bardet–Biedl syndrome, Joubert syndrome, Meckel–Gruber syndrome, and a variety of other diseases that affect the kidney, heart, brain, liver, and the muscular-skeletal system, among other organs and systems (Youn and Han 2018).

The primary cilium has a similar structure to that of the better-known hair-like cilia and flagella whose functions have been well-documented as being motile tail-like appendages, either involved in cell locomotion or in the movement of fluids outside the cell surface. However, unlike its motile ciliary brethren, the primary cilium is involved in sensory function rather than motility. Primary cilia are comprised of a barrel-like structure called an axoneme that extends from an organelle known as the basal body (Figure 18.1, see schematic diagram and micrograph where the green meets the red), which is essentially a mother centriole, the older of the two centrioles that make up the centrosome. The centrosome itself is a key cellular organelle required for cell division but it may be "repurposed" for ciliogenesis in non-dividing (non-mitotic) cells by moving away from the cell center and establishing the primary cilium. The ciliary axoneme that extends toward the cell membrane typically contains nine pairs of microtubule doublets and is surrounded by a specialized membrane (see Figure 18.1). Until recently there were competing hypotheses regarding the functional role of the primary cilium: (1) a vestigial organelle with little

DOI: 10.1201/9781003202974-18

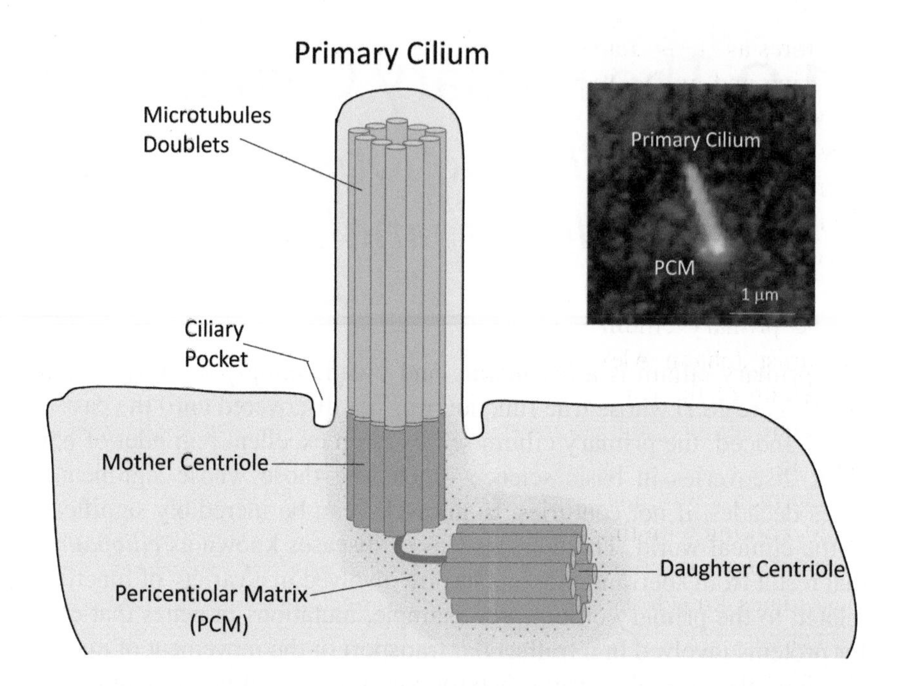

FIGURE 18.1 The primary cilium. Schematic diagram illustrating the structure of the primary cilium. The mother and daughter centrioles and pericentriolar matrix that comprise the centrosome move toward the plasma membrane. The mother centriole is the basal body from which the microtubule doublets extend, surrounded by a ciliary membrane that fuses with the cell surface plasma membrane and extends into the extracellular space to form the primary cilium. The inset on the right shows an immunofluorescence micrograph of a primary cilium in human retinal pigment epithelial cells. These cells were fixed and immunostained with primary and secondary antibodies to mark the primary cilium (with acetylated tubulin; shown in green) and with anti-γ-tubulin antibodies to mark the pericentriolar matrix (PCM) region (shown in red). Image provided courtesy of the author.

or no remaining physiological significance, (2) an organelle that serves to sequester the centriole so as to prevent spindle formation, chromosome segregation between mother and daughter cells, and cell division, and (3) a sensory organelle involved in signal transduction events (Satir, Pedersen, and Christensen 2010). While today there is little doubt that the major function of the primary cilium is related to its sensory role as a signaling organelle, for many years researchers had difficulties letting go of the predisposed notion that all cilia—including the primary cilium—were involved in some form of motility.

The great microscopist Antonie van Leeuwenhoek was probably among the first to observe cilia as the bacterial flagellum, and he described these structures as "tails" found in the "divers animals" in rain water that had a "pleasing and nimble motion" and were "often tumbling about and sideways" (Leeuwenhoek 1677). However, the immotile primary cilium was only described many years later. Given that the primary cilium is in a sense a centriolar extension, it is not surprising that the discovery of the primary cilium came on the heels of intense studies of the centrosome, which was discovered independently by Walther Flemming, Edouard Van Beneden, and Theodor Boveri in the late 1800s (Alvarado-Kristensson 2020).

The primary cilium itself was apparently first depicted by German anatomist Johann Alexander Ecker in 1844, who was working on the epithelium of the ear canals of the sea lamprey and observed that each epithelial cell contained a single cilium (Ecker 1844, Bloodgood 2009). Both Russian embryologist and anatomist Alexander Kowalevsky and the German Paul Langerhans (who had also discovered Langerhans cells while working with the famous pathologist, Rudolph Virchow) reportedly depicted primary cilia in publications studying the small, eel-like cephalochordates known as *Amphioxus* (Bloodgood 2009, Kowalevsky 1877, Langerhans 1876). However, credit for first identifying the primary cilium has often been awarded to Karl Zimmermann at the University of Berne in Switzerland. Zimmermann was the first to observe these organelles in mammalian cells, and he coined the term *centralgeissel* or central flagella (Zimmermann 1898, Bloodgood 2010). Remarkably, Zimmermann also pointed out that he never observed any movement of the primary cilium, which directly contrasted with the known function of all other cilia described to date (as well as the name he gave them), foreshadowing our current knowledge of primary ciliary function.

Zimmermann is usually credited with the discovery of the primary cilium. Although this is not an entirely accurate assignation of credit, the world-at-large often prefers an easily attributable "hero" as opposed to the more convoluted type of credit that is often more apt for scientific discovery. In any case, Zimmermann's highly significant observations have held up to the test of time. Zimmerman clearly understood the intimate relationship between the primary cilium and centrosomes; more specifically, he realized that the primary cilium always extended from the older of the centriole pair, the mother centriole whose localization is farthest from the nucleus and closest to the plasma membrane.

The initial discovery of the primary cilium, whether it is credited to Zimmerman, or arguably to Flemming, Van Beneden, and Boveri, had a profound and possibly unexpected impact on the fundamental understanding of cell biology. Indeed, in the late 1800s, just prior to the turn of the century, two papers by established histologists and cell biologists led to a revolutionary new hypothesis that has changed how cell biologists view the cell. Hungarian Mihaly Lenhossek and French scientist Louis Félix Henneguy separately published papers providing evidence that the very same centriole at the cell center that generates the mitotic spindles required for cell division is also the so-called basal body that sits at the base of the primary cilium as it extends its axoneme outward to the cell surface. These studies together led to the long-standing hypothesis that is known as the *Lenhossek–Henneguy hypothesis*: since these structures (the basal body from which the cilium extends, and the centriole at the cell center involved in mitotic spindle formation) are one and the same, this means that *ciliated cells* (cells that have generated a primary cilium) must therefore *lack a centralized centrosome* (since the basal body from which the cilium extends moves toward the plasma membrane in the course of ciliogenesis) and *are thus unable to divide* (Chapman 1998).

There was much experimental evidence in favor of the Lenhossek–Henneguy hypothesis, and many of the scientists who studied centrosomes early in the twentieth century supported the idea. Despite this, there was a degree of opposition from those who claimed they had observed cell division in ciliated cells and the hypothesis was not yet unanimously accepted. Ironically, it was later demonstrated that the "reportedly-ciliated" dividing cells did not actually have primary cilia. In the 1950s, various studies were published that showed that centrioles and basal bodies were structurally the same (Burgos and Fawcett 1955, deHarven 1956), with the "icing on the cake" in support of the Lenhossek–Henneguy hypothesis coming from quite a recent study showing that a centriolar protein called HYLS-1 is incorporated into the outer centriolar wall, but has no direct function in centriole assembly (Dammermann et al. 2009). Indeed, its role was found relegated exclusively to ciliogenesis, with a single amino acid substitution in the protein causing the development of a severe ciliopathy and thus highlighting the dual function of centrioles (Dammermann et al. 2009).

As alluded to earlier, it is interesting that for many years and well into the twentieth century, the primary cilium was known as the *central flagellum*, suggesting, of course, a motile function. E.V. Cowdry at Rockefeller University in New York published a study in 1921 on "Flagellated Thyroid Cells in the Dogfish" in which he essentially showed no evidence that the primary cilia in the thyroids of these organisms were involved in motility in any way—and yet, consistent with the difficulties scientists have in letting go of old dogma, he could not bring himself to clearly support the notion that the primary cilium had another function (Cowdry 1921). Indeed, unable to completely abandon the idea of motility, he wrote:

> It is almost inconceivable that such an elaborate mechanism should not serve some useful purpose. Although I have not actually observed

> movement of the flagella, they have all the distinctive characters of a typical motor apparatus such as one meets within the protozoa or in hydra.

However, completely lacking any evidence to support a motile function (and being unaware of any other possible function), Cowdry then suggested that the primary cilium might merely be a vestigial organelle that is an evolutionary remnant that nonetheless serves no useful purpose (Bloodgood 2009, Cowdry 1921).

With the advent of transmission electron microscopy in the middle of the twentieth century, a number of researchers turned back to the *central flagellum*, among them Sergei Sorokin at Harvard, who pioneered many of the key morphological studies and coined the term "*primary cilia*," although he sometimes called them *rudimentary cilia* indicating that their true function was still not understood (Sorokin 1962). Indeed, it was not until the turn of the twenty-first century that a series of researchers began to find evidence that the primary cilium is indeed a sensory organelle—an idea floated over 100 years earlier by Zimmermann (Pazour and Witman 2003). In parallel, the increasing number of new ciliopathies identified affecting so many different organs highlighted the need for the primary cilium to be seriously addressed, and not regarded merely as a rudimentary organelle (Youn and Han 2018).

From the discovery of the primary cilium to the understanding of its function as a sensory organelle and its role in human disease, the last ~150 years of research on the primary cilium mirror many of the great discoveries in science. Poor communication (perhaps even exacerbated by the language barriers entailed in the published literature), the reluctance of researchers to abandon scientific doctrine even in the light of clear evidence, and the slow, cumulative collection of new evidence are easily identifiable in the discovery of the primary cilium, as well as with many of the other great biomedical discoveries. Ultimately, the discovery of the primary cilium and an understanding of its role in the cell may lead to new medical treatments for the vast number of ciliopathies caused by defective proteins. While it may not be possible to "correct" developmental problems, the coupling of fundamental knowledge of genetic engineering and CRISPR technologies with the knowledge of defective proteins inherent in specific ciliopathies may potentially lead to new treatments.

19 The Discovery of the Golgi Complex

*A Pivotal Organelle with Multiple Functions***

Also known as the Golgi apparatus, the discovery of the Golgi complex showcases a classic example of the nature by which many great scientific discoveries are made. The initial discovery of the Golgi apparatus by Italian scientist Camillo Golgi in 1898 (Golgi 1989a, b) came directly on the heels of acceptance of "Cell Theory"—the notion that cells are the primary building blocks of tissues and that individual cells are comprised of three major components: (1) a nucleus, (2) fluid, and (3) a surrounding wall or membrane (Wolpert 1995). Cell theory, ultimately relating back to the microscopy work of Robert Hooke who famously noted that "by the help of microscopes, there is nothing so small, as to escape of enquiry" (Hooke 1665), was promoted by two scientists in Berlin, Germany, in the mid-1800s, who trained with Johannes Mueller: Matthias Jacob Schleiden and Theodor Schwann. However, it was only 40–50 years later when scientists first began to understand that the contents of cells extended well beyond a mere sac of fluid that the first intracellular structures (aside from that of the nucleus) were observed. It was during this phase that Camillo Golgi first identified the structure that has retained his name to this day.

When Golgi published his findings on the existence of a series of reticular-shaped *intracellular canals* in Purkinje cells as well as cells from spinal ganglia (Golgi 1989a, b), his studies were met with opposition by other researchers who maintained that his discoveries resulted from artifacts of the methodology by which he "impregnated" or stained his cells for visualization. Indeed, many scientists at the time claimed that the methods used by Golgi caused the retention of artificial depositions of heavy metals within the cells, thus distorting and invalidating any conclusions drawn. Despite additional data that surfaced suggesting that these reticular structures observed deep within the cell were visible in nearly every cell type tested (albeit with some morphological variability), because the

DOI: 10.1201/9781003202974-19

studies were all performed using Golgi's controversial impregnation methods, opposition to the actual existence of this Golgi structure was not put to rest until the advent of new electron microscopy methods in the 1950s. The discovery of the Golgi apparatus again highlights another key scientific finding whose significance and complete understanding awaited the development of technologies that could be used to better study it.

In 1954, a study by Albert Dalton and Marie Felix at the National Institutes of Health in Bethesda, Maryland, clearly showed that the Golgi apparatus is comprised of a series of vacuoles, lamellae, and vesicles (Figure 19.1; see schematic diagram and fluorescence micrograph) leading them to coin the term "Golgi complex" (Dalton and Felix 1954). Based on their studies using both the Golgi metallic impregnation methods as well as light microscopy on many diverse cell types, they also concluded that

> it is reasonable to consider the Golgi complex a distinct morphologic entity and to classify it as one of the cytoplasmic organelles along with plasma membranes, mitochondria, ergastoplasm (author's note: this was the term used to describe the endoplasmic reticulum, or ER), and centrioles. Its designation as an organelle is not considered as being incompatible with the existence of an intimate relationship between the complex and other cell components.
>
> **(Dalton and Felix 1956)**

As described in a review by two scientists who played crucial roles in the emerging ideas and research on the Golgi complex, Marilyn Farquhar and George Palade (who also happened to be married to one another), the studies of Dalton and Felix were the turning point in our modern understanding of this organelle, and were soon followed by additional support for its existence—and subsequent comprehension of its function (Farquhar and Palade 1981). However, despite the validation of the existence of the Golgi complex, and the emergence of new technologies to probe its composition, Farquhar and Palade observed that "Information on the function of the Golgi complex was limited, however, to noting the topographical association between this organelle and forming secretion granules" (Farquhar and Palade 1981).

Camillo Golgi was apparently reluctant to speculate too much on the potential function of his new organelle as noted by Marina Bentivoglio (Bentivoglio 1999). However, rather prophetically he noticed the altered distribution and change in form in mucous glands of the frog stomach of the reticular apparatus in relation to the cell's secretory activity, suggesting a role for the Golgi complex in secretion (Bentivoglio 1999). Moreover, in 1923 Robert Bowen at Columbia University in New York published a

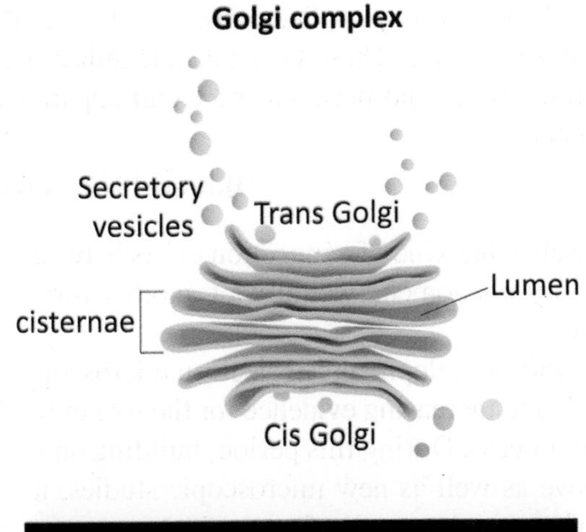

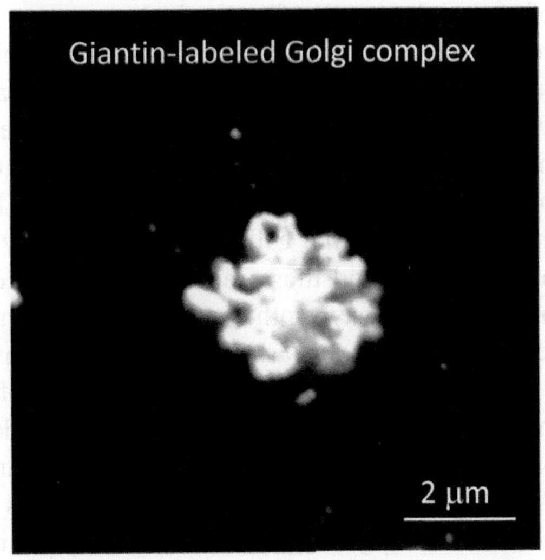

FIGURE 19.1 The Golgi complex. Schematic diagram illustrating the structure of the Golgi complex, showing the cis Golgi that receives incoming newly synthesized proteins from the endoplasmic reticulum, the cisternae or Golgi stacks, and the trans Golgi that releases secretory vesicles for transport to the plasma membrane. Below the illustration is a fluorescence microscopy image of the Golgi complex. Human HeLa cells were fixed and immunostained with an antibody that recognizes the Golgi-localized protein, Giantin. Image provided courtesy of the author.

paper entitled "The Origin of Secretory Granules" in which he references his own work and that of D.N. Nassonov and notes that

> There is abundant evidence from the results of both Nassonov and myself, that the granules on leaving the Golgi apparatus actually are provided with a miniature replica of the Golgi material in the form of

> a small cap, which is closely applied to the periphery of the granule-a point of great importance. These facts taken together clearly indicate a connection of some kind between the Golgi apparatus and secretory phenomena.
>
> **(Bowen 1923, Nassonov 1923)**

However, precisely what kind of connection exists between the Golgi complex and secretory phenomena would wait another three decades before being unraveled.

By the mid-1950s, developments in electron microscopy as well as other methodologies led to increasing evidence for the role of the Golgi complex in the secretion process. During this period, building on the prior observations noted above as well as new microscopic studies, it was established that the type of material typically found within secretory granules could also be observed in the Golgi complex (Farquhar and Wellings 1957). In addition, the new technique of *cell fractionation*—the separation of intracellular organelles through centrifugation at variable speeds—was being used for biochemical analysis of the Golgi, and it was combined with another exciting new method known as *autoradiography*. For autoradiography, researchers fed tissues and cells with radioactively labeled amino acids for a period of time, and these radioactively labeled residues then became incorporated into newly synthesized proteins (in the same manner that non-labeled amino acids are assembled into proteins), thus "marking" them. By labeling new proteins for a select period of time and then removing from the cells all of the extra radioactively labeled amino acids that *did not* assemble into new proteins (to stop the labeling of any more proteins that that time point onward), a method known as *pulse-chase*, it was then possible to monitor the whereabouts and dynamic movement of the radioactively labeled proteins as they were transported from organelle to organelle within the cell. By coupling fractionation and autoradiography, it was thus possible to obtain clear evidence that "secretory proteins are transported from the cisternae of the rough endoplasmic reticulum to condensing vacuoles of the Golgi complex via small vesicles located in the periphery of the complex" (Jamieson and Palade 1967). This conclusion was a key one that helped implicate the Golgi complex in the secretory process, by being among the first studies to functionally (rather than exclusively morphologically) link the Golgi complex to secretory function.

With the growing evidence that the Golgi complex was required for the packaging (and in many cases, the concentration) of secretory proteins in vesicles bound for transport to the cell surface, biomedical researchers began to further address the role of this unique organelle. It became clear that the Golgi complex is the site of synthesis and attachment of complex carbohydrates to proteins (Neutra and Leblond 1966). In addition, a novel function for the Golgi complex emerged in the "tagging" of lysosomal hydrolases with a molecule known as mannose-6-phosphate. Indeed, this post-translational modification of the lysosomal enzymes through the covalent attachment of a mannose-6-phosphate moiety to them leads to their recognition by a specific receptor localized to the Golgi complex known as mannose-6-phosphate receptor. The binding of the mannose-6-phosphate-containing enzyme at the Golgi complex by mannose-6-phosphate receptor is required for the packaging of these lysosomal enzymes into transport vesicles at the trans Golgi network (TGN) for their delivery to late endosomes and lysosomes (Kornfeld 2018). Accordingly, the Golgi complex, in addition to its critical function in supporting the secretion of proteins, is also responsible for what has been termed as *biosynthetic transport*, or the trafficking of newly synthesized proteins that function and reside in endocytic organelles such as lysosomes and endosomes (Figure 19.2).

Given the significance of the Golgi complex in secretion and biosynthetic transport within the endocytic system, it is not surprising that a large number of diseases have been identified in which the Golgi complex (or

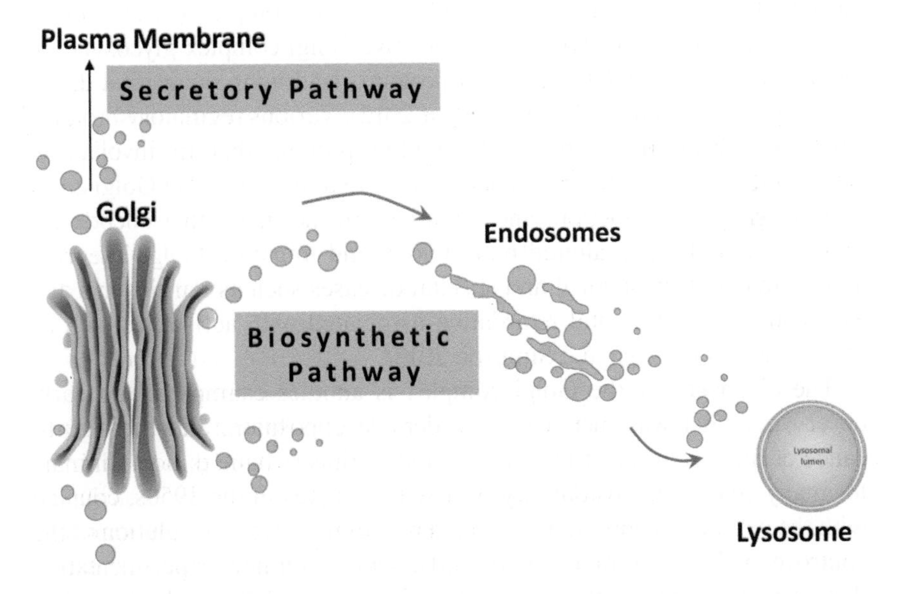

FIGURE 19.2 Key functions of the Golgi complex. The Golgi complex is the major secretory organelle that facilitates secretion of proteins and targets newly synthesized receptors to the cell surface membrane (*secretory pathway*). However, the Golgi complex is also a sorting station that controls the delivery of newly synthesized resident proteins of the endosomal system to endosomes and lysosomes (*biosynthetic pathway*).

proteins that localize to the Golgi complex) is involved. Since the Golgi complex is essential for a wide number of post-translational modifications to proteins that are trafficked through it, including various forms of glycosylation (addition of a sugar group), sulfation (addition of a sulfate group), and proteolytic trimming that occurs as the proteins traverse the Golgi stacks, mutations that affect the function of the Golgi complex would be anticipated to be disruptive to the cell. Indeed, a variety of mutations in proteins that interfere with general trafficking to and from the Golgi complex have been identified that cause neurodegenerative diseases. One example is the proteolipid protein, PLP1, which when mutated, misfolds and accumulates in the endoplasmic reticulum, thus upsetting the balance of transport to and from the Golgi complex and the endoplasmic reticulum. This impairment in membrane trafficking ultimately causes the Golgi complex to fragment and it is assumed that this has a bearing on the abnormal myelin development observed in the nervous system that occurs in Perlizaeus–Merzbacher disease (Inoue 2019). In a disorder of connective tissue known as *Cutis laxa*, mutations in the *ATP6V0A2* gene that encodes an ATPase subunit lead to inefficient glycosylation of various serum proteins that normally occurs within the Golgi complex, as well as impaired movement of various proteins through the Golgi complex (Fischer et al. 2012). Other mutations that lead to defective Golgi complex glycosylation include those in the COG family of proteins, leading to congenital disorders of glycosylation (Miller and Ungar 2012). Various regulatory proteins from the Rab family of small GTP-binding proteins that are involved in membrane trafficking have also been implicated in controlling Golgi complex integrity and function, and mutations in the genes that encode for these proteins lead to altered morphology and impaired Golgi integrity, and ultimately neurological and skeletal diseases such as Parkinson's disease and X-linked mental retardation associated with autism, as well as other illnesses (Bexiga and Simpson 2013).

The discovery of the Golgi complex is another example of an early observation that was met with considerable opposition, and the significance of the discovery of the Golgi complex thus remained indeterminate for many years. The availability of new techniques in the 1950s, coupled with the work of curious and motivated researchers, led to resolution of the controversy, but even more importantly, spurred on new experimentation that helped determine the crucial functional role of the Golgi complex. Although specific cures for the many Golgi-related diseases have yet to surface, one can easily envision that modern molecular biology techniques coupled with a clear understanding of the role of the Golgi and its affiliated proteins will ultimately lead to new possibilities for disease treatment in the coming years.

20 The Lysosome

*A Trash Bin and End of the Road for Many Cellular Molecules***

The lysosome is a unique membrane-bound organelle found in eukaryotic cells whose primary function is to degrade both intracellular and extracellular materials that are taken up into cells (Caplan 2003). While lysosomes are not easily distinguishable morphologically from other endocytic compartments such as endosomes, they are characterized by having about 50 hydrolytic enzymes that function at the acidic pH which is typical within the limiting membrane of the lysosome, known as the lysosomal lumen (Figure 20.1; see schematic diagram and fluorescence micrograph). These enzymes are essential for breaking down proteins, lipids, DNA, RNA, sugars, and other cellular building blocks, so that new proteins and macromolecules can be synthesized from the recycled components. Despite carrying out such incredibly important degradative functions that when impaired can cause well over 50 different diseases (many are known as *lysosomal storage disorders* due to the build-up of materials that are normally degraded in the lysosomes of healthy individuals), the identification of the lysosome as an independent organelle did not occur until its rather serendipitous discovery by Christian de Duve and his co-workers in the mid-1950s. Indeed, de Duve (at the Catholic University of Leuven in Belgium and Rockefeller Institute in New York, NY), Albert Claude (at the Free University of Brussels and Catholic University of Leuven in Belgium, and Rockefeller Institute in New York, NY), and George Palade (at the Rockefeller Institute in New York, NY, Yale University in New Haven, CT, and the University of California, San Diego) would share a Nobel Prize for their findings in 1974.

In some ways, the discovery of the lysosome is reminiscent of the discovery of insulin (as detailed in Chapter 11), about 30 years earlier: "All we wanted was to know something about the localization of the enzyme glucose 6-phosphatase, which we thought might provide a possible clue to the mechanism of action, or lack of action, of insulin on the liver cell." In his Nobel Prize lecture that was published in *Science*, de Duve described

DOI: 10.1201/9781003202974-20

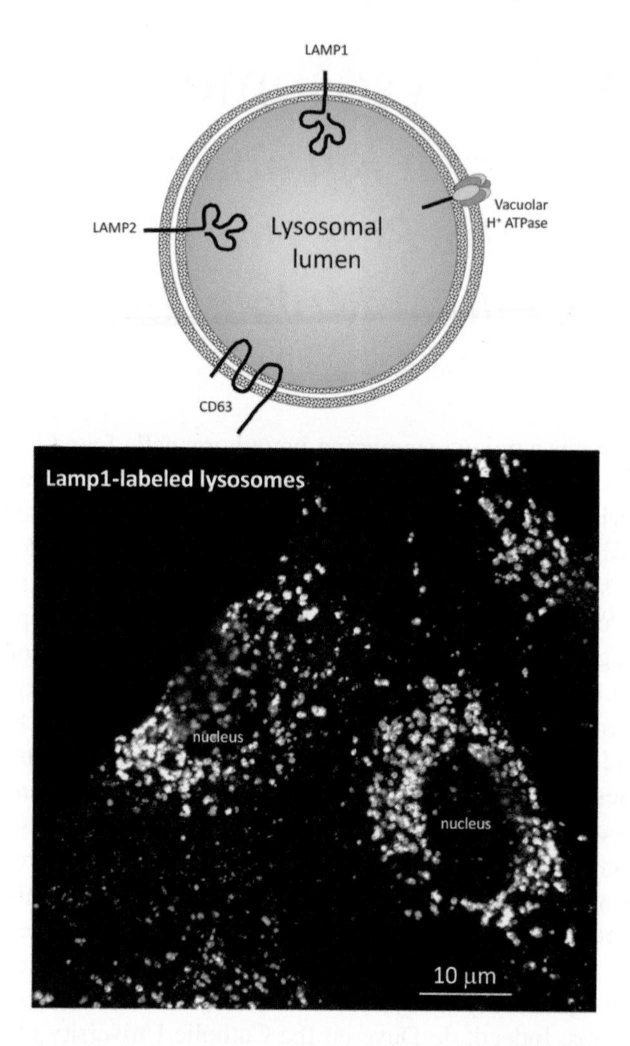

FIGURE 20.1 The lysosome. Schematic diagram showing the lysosome with its limiting membrane and membrane proteins, such as LAMP1 and LAMP2 and CD63, as well as the vacuolar ATPase protein pump which maintains an acidic pH in the lysosomal lumen, where the lysosomal hydrolases carry out degradative activity. Below is a fluorescence microscopy image of lysosomes in the cell. Human HeLa cells were fixed and immunostained with an antibody that recognizes the lysosomal membrane protein LAMP1. Image provided courtesy of the author.

his use of centrifugation to roam through cells as being similar to the flight of a boy over Sweden on the wings of a white goose in Selma Lagerlöf's classic *Nils Holgersson's Wonderful Journey across Sweden* (Duve 1975).

Using centrifugation as a crude means to separate nuclei, mitochondria, and microsomal fractions, de Duve and colleagues realized that a series of hydrolytic enzymes including acid phosphatase, cathepsin, ribonuclease, and deoxyribonuclease all appeared to localize to the same fractions, and by combining data from various experiments they were confident enough to propose the existence of a new organelle, the lysosome.

In a paper detailing the discovery of the lysosome, Dorothy Bainton highlights the serendipity of this finding by imagining how such a study would have been cast in terms of a modern-day grant proposal and progress report. Specific Aims are the backbone of a scientific grant proposal, outlining the main goals, whereas the Progress Report details the advances made once a grant proposal has been awarded, usually on a yearly basis. Bainton creatively invents the following proposal and subsequent progress:

> Specific aim: to localize the enzyme glucose-6-phosphatase. Significance: to elucidate the mechanism of action of insulin on the liver. 1952: Progress Report: Unfortunately, no progress has been made on this problem; rather, we would like to report on . . . "From Insulin to Latent Acid Phosphatase".
>
> **(Bainton 1981)**

Bainton points out that despite clear plans to localize the enzyme, the researchers hit upon something of interest (as described in the imaginary progress report) and were astute enough to realize that they had made a highly significant discovery that rendered their original goal trivial in comparison.

The lysosome has become such a household name in the biomedical sciences that it is easy to gloss over the enormous ramifications of the discovery of this organelle. For a start, while the lysosome is the key degradative organelle in most cell types, many other types of cells have *lysosome-related organelles* (LRO) that play a dual role in both degradation and in additional and more specialized functions (Dell'Angelica et al. 2000). An example of an LRO is the melanosome, which serves as a pigment organelle in melanocytes (a type of skin cell) in addition to its role in degradation. In addition, lytic granules play a dual role in T lymphocyte cells in degradation and are also released from the lymphocytes to kill target cells identified by the immune system (Dell'Angelica et al. 2000). Another example is the Major Histocompatibility Class II compartment, which in addition to playing a typical lysosomal-like degradative role is also used in the presentation of antigens by B cells and antigen presenting cells (Dell'Angelica et al. 2000, Caplan et al. 2000). Intriguingly, there are dozens of known diseases that are related to lysosomal dysfunction, and many of them are *autosomal recessive* diseases (passed down hereditarily by healthy parents who each, often unknowingly, carry a recessive gene). As noted, these diseases are known collectively as lysosomal storage disorders, and have an estimated combined incidence of about 1:8,000 births. Typically, these diseases result from a deficiency of a specific lysosomal enzyme that is either mutated, poorly expressed in the cells, misfolded, and/or dysfunctional and otherwise unable to degrade a specific substrate, thus resulting in the subsequent accumulation of high levels of the substrate that cause cellular toxicity. Such is the case in Tay–Sachs disease, in which the β-hexosaminidase enzyme subunit is deficient, leading to excessive accumulation of the sphingolipid ganglioside GM2, and progressive neurological degeneration (Solovyeva et al. 2018). Why excessive lysosomal storage appears to be especially detrimental to neurons is not altogether clear, although one might speculate that neuronal function is particularly sensitive to the build-up of some substrates. Fortunately, however, due to advances in understanding the cause of lysosomal storage diseases such as Tay–Sachs, and the advances in molecular biology that allow PCR-based testing of amniotic fluid, prospective higher-risk parents (for example, those with a Jewish Ashkenazi background who are more frequently heterozygote carriers with a mutant recessive allele) can screen for this horrific disease at an early stage.

Until the last decade or so, the only potential treatment for most lysosomal storage diseases was a hematopoietic stem cell transplantation (see Chapter 13). The idea with this treatment was to provide new donor cells that contain the missing enzyme, with the hope that the stem cells infiltrate into the various tissues and organs of the recipient, and release the enzyme—allowing it to be taken up by the recipient's cells and used to catabolize whatever substrate is building up and causing damage. As one might envision, however, such a process is hard to control, incomplete, and slow, and unlikely to really provide a full solution to the disease.

Recently, researchers and doctors have developed a new strategy called *enzyme replacement therapy* (ERT) as a novel method to treat many of the lysosomal storage disorders that are caused by a missing enzyme. Over a dozen different ERT treatments have been approved to address illnesses such as Gaucher disease, Niemann–Pick disease, Fabry disease, and many others. The overall idea is fairly simple—the missing enzyme is synthesized *in vitro* (through technologies that arose beginning with the solution of the genetic code), and then applied intravenously to the patient. The injected enzyme finds its way through the blood serum and then encounters and binds to a receptor on the cell surface known as mannose-6-phosphate receptor, which normally is used to transport the enzymes from the Golgi following their synthesis to the lysosomal lumen where they reside and carry out their functions (Neufeld 2011). However, there are also levels of mannose-6-phosphate receptor that are localized to the cell surface membrane, and in the case of ERT, they can bind to the missing enzyme when it is in the serum and direct it into the cell and onward to the lysosome, thus bypassing the need for endogenous enzyme in the cell. For a longer-term solution, the corrected expression of defective enzymes that cause lysosomal storage diseases will likely be targets for the use of CRISPR technology in the coming years.

Intriguingly, in some cases the lysosome can be utilized by parasites that cause infections and disease states; indeed, the lysosome can be a protected environment for certain intracellular pathogens, such as *Mycobacterium tuberculosis* (Hart 1968). In addition, lysosomal function is required for the life cycle of the malarial parasite (Homewood et al. 1972). Ultimately this has led to the discovery of various inhibitors for lysosomal function, including the drug chloroquine, which is widely used for malaria prophylaxis, although ultimately the mechanism by which it inhibits the malarial parasite is still poorly understood.

There is no doubt that in this sense, the discovery of the lysosome again shows the significance of a well-defined research plan for scientists. While there is a need for a specified and clear plan of action, at the same time there is also a need for scientists to be able to shed pre-conceived notions and display flexibility, and to be willing to go off the beaten path in search of answers to new challenges and questions that arise unexpectedly in the course of research. The discovery of the lysosome again highlights the way that technological and conceptual advances must move forward hand in hand. Without powerful light microscopy techniques at the time to resolve intracellular organelles, lysosomes were not easy to discover. The power of sedimentation studies, as described by de Duve who avidly read the book written by Svedberg and Pedersen (Svedberg 1940), resolved some of the key technological advances that limited and perhaps delayed the discovery of the lysosome. Moreover, in addition to its clinical importance (as highlighted above), the discovery of the lysosome undoubtedly paved the way for other significant biomedical discoveries, such as the receptor-mediated endocytosis (see Chapter 22).

21 The Ubiquitin-Proteasomal Pathway
*Targeted Protein Degradation and More****

In 1975, a short paper by David Schlesinger and Gideon Goldstein of the Memorial Sloan-Kettering Cancer Center in New York, NY, appeared in the journal *Nature* (Schlesinger and Goldstein 1975). By today's standards, this report was relatively modest in its findings, essentially using classic amino acid sequencing techniques to decipher the full 74 amino acid sequence of a protein called ubiquitin. This protein, which was originally known as ubiquitous immunopoietic polypeptide (UBIP), was initially observed in the thymus of cows, but soon after it was found in all animal cells, as well as yeast, higher plants, and even bacteria. Remarkably, the authors wrote that "although its physiological function remains unknown, ubiquitin must be vital to the living cell to have been conserved over such a long evolutionary time span" (Schlesinger and Goldstein 1975). This statement turned out to be no less than prophetic; in 2004 the Nobel Prize for Chemistry was awarded to three researchers, Avram Hershko and Aaron Ciechanover of the Technion Institute in Haifa, Israel, and Irwin Rose (who was at the time at the University of California, Irvine), for their discovery of the ubiquitin-mediated proteasomal degradation pathway.

Just a few years after the discovery of the protein termed "ubiquitin," Hershko and his co-workers were busy challenging a long-held assumption that most cellular proteins were typically long-lived—that is, that once translated from mRNA into proteins, the proteins existed in the cell for long periods of time prior to undergoing degradation. In a series of seminal studies, they demonstrated that in a cell-free system (with the components derived from rabbit reticulocytes) a protein designated as ATP-dependent proteolysis factor-1 (APF-1) was found covalently bound to multiple protein substrates that underwent proteolytic degradation. Their conclusion was that APF-1 was likely marking these substrate proteins as targets for a protease, and channeling them to undergo degradation (Ciechanover et al. 1980). In other words, Ciechanover and Hershko had obtained evidence

DOI: 10.1201/9781003202974-21

that this APF-1 protein might be part of a system or cascade that first marks proteins and then targets them for degradation in a pathway independent of lysosomes. Subsequent studies indicated that this proteasomal degradation system required a series of three enzymes, acting sequentially to bring the APF-1 protein to the substrate target and induce its covalent attachment to the target (Hershko and Ciechanover 1982). Remarkably, it was quickly discovered that APF-1 was none other than the protein already identified as ubiquitin (Wilkinson, Urban, and Haas 1980), sparking a new era in the fundamental understanding of protein degradation, and initiating research on the ubiquitin-proteasomal degradation pathway.

In the ensuing decades, great advances were made in understanding this novel mechanism for the dispatching of "old" proteins and their subsequent degradation. It was quickly demonstrated that such a degradation pathway exists in living cells, and is not merely an interesting effect observed *in vitro*, exclusively in cell-free test tube systems (Finley, Ciechanover, and Varshavsky 1984, Ciechanover, Finley, and Varshavsky 1984). Essential details were worked out regarding the ubiquitin degradation pathway, including the determination that protein degradation was induced not by the linkage of a single ubiquitin protein to a target substrate, but rather by an entire chain of ubiquitin proteins, with the first binding to the target substrate, and then subsequent ubiquitin proteins each binding to the preceding one, forming a *polyubiquitin chain* (Chau et al. 1989). In addition, Alexander Varshavsky and his colleagues discovered the famous *N-end Rule*, formulating the idea that a protein's longevity within the cell directly depends upon the amino acid residues at the N-terminus of the protein (known as a *degron*), which ultimately bind the ubiquitin ligase enzyme and support the latter's ability to covalently couple a ubiquitin to it and effectively induce its degradation (Bartel, Wunning, and Varshavsky 1990).

So how does the ubiquitin degradation pathway actually work? Through decades of work by hundreds of laboratories, it is now known that a cascade of three enzymes, called E1, E2, and E3, act sequentially to activate, carry, and attach a ubiquitin protein to a substrate protein (Figure 21.1). Successive addition of ubiquitins to one another through the binding of one ubiquitin protein to a lysine residue on the previously attached ubiquitin protein (*polyubiquitination*) ultimately leads to a chain of ubiquitins that represents a signal for degradation that is recognized by a cytoplasmic organelle comprised of a very large complex of proteins known as the proteasome (Waxman, Fagan, and Goldberg 1987). Through the energy released upon ATP hydrolysis, polyubiquitinated proteins are dispatched through the proteasome where they undergo fragmentation into small peptides and individual amino acids. Thus, ubiquitin-dependent degradation provides an alternative mechanism to lysosomes for the degradation of

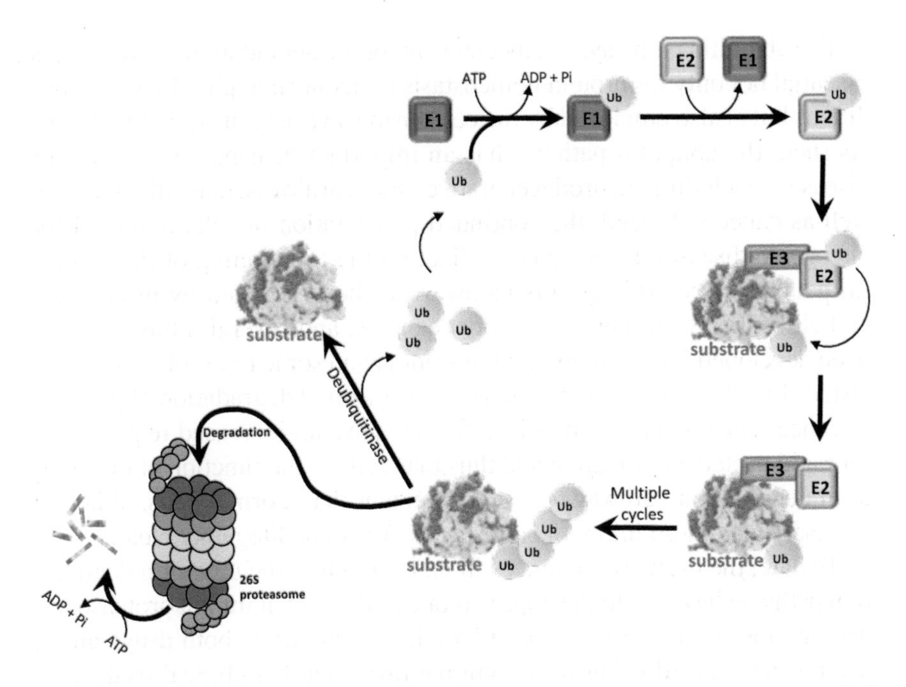

FIGURE 21.1 The ubiquitin-proteasomal degradation pathway. The small 8 kilodalton ubiquitin protein (Ub; in pink) is carried and transferred by an enzyme called E1 to a second enzyme called E2. The E2 enzyme either transfers the ubiquitin to an E3 enzyme, or functions together with the E3 enzyme to covalently attach the ubiquitin to the substrate protein that is slated to undergo degradation via the proteasome. Once a substrate has been ubiquitinated, additional ubiquitin proteins are then attached serially to the preceding ubiquitin, forming a polyubiquitin chain. A polyubiquitinated substrate, if it does not undergo deubiquitination by a deubiquitinase enzyme that removes the ubiquitins, is ultimately targeted to the proteasome where it is cleaved and hydrolyzed in an ATP-dependent process.

proteins, a pathway that can be honed and acutely fine-tuned to regulate critical cellular events.

Despite the obvious fundamental implications of understanding the mechanisms by which cellular proteins are degraded, and how this process is controlled, perhaps one of the most significant discoveries related to the ubiquitin degradation pathway came in the early 1990s, when two groups, including that of Hershko, showed that a family of proteins known as cyclins that serve as key regulators in cell cycle and division are substrates for ubiquitin and undergo degradation as a result of ubiquitination (Glotzer, Murray, and Kirschner 1991, Hershko et al. 1991). This finding has immense implications, as it suggests that control of the ubiquitin degradation pathway might allow researchers and physicians to influence cell cycle, with obvious significance for cancer.

The ubiquitin pathway, as a mechanism for the degradation of proteins, is essential not only for protein homeostasis under normal physiological conditions but is also employed by the cell during a variety of stress responses. As such, the ubiquitin pathway has an important bearing on a variety of diseases, including neurodegenerative and cardiovascular illnesses, as well as cancers. Indeed, the continued proliferation of cells (required for oncogenic disease) is very much reliant on the functioning of the ubiquitin pathway. Accordingly, this pathway has been targeted by researchers and clinicians for inhibition in select cancers, and several inhibitors have been developed for use in multiple myeloma and some types of lymphoma. Ostensibly, the rationale for inhibiting proteasomal degradation of proteins in cancer cells is that neoplastic cells may have an increased requirement for rapid protein degradation and thus may rely on the function of the ubiquitin degradation pathway to a greater extent than normal cells, although precisely why these inhibitors are successful is not altogether clear.

To date, the dramatic advances in understanding the ubiquitin degradation pathway have led to the approval of two drugs that inhibit proteasomal degradation: bortezomib and carfilzomib. At this time, both drugs are in use for patients suffering from multiple myeloma, but clinical trials have recently been completed for use in combination with other drugs in a variety of other cancers, including biliary tract cancers, breast cancers, head and neck cancers, various lymphomas, and many other neoplastic diseases (Obrist et al. 2015). Moreover, a variety of newer *third-generation* proteasomal inhibitors are currently being tested in different stages of clinical trials, with the hope that some will prove helpful. Thus a mere 43 years have passed since Schlesinger and Goldstein first identified the protein ubiquitin, and around 30 years have passed from the time Ciechanover, Hershko, and their colleagues first began to decipher the role of ubiquitin and its remarkable pathway in the degradation of cellular proteins until this information could initially be converted or *translated* into findings with clinical ramifications for treating patients. This highlights the need for today's basic science, which may not intuitively be valued as leading to cures, but within the next 25–40 years may well provide the next generation of successful treatments for cancer and other diseases.

22 Receptor-Mediated Endocytosis

*Gateway to the Cell***

All cells carry out an essential function known as *receptor-mediated endocytosis*. When specific receptors localized to the cell surface membrane (also known as the plasma membrane) engage and bind to protein ligands in the extracellular milieu, they subsequently transmit signals into the cell through a series of biochemical reactions. Just as important for the cell is the ability to shut down this signal transduction cascade, to avoid, for example, over-stimulation that might lead to incessant proliferation and cancerous behavior. One way to prevent such over-stimulation at the cell surface is through a process of *internalization*, another term for receptor-mediated endocytosis. During this internalization process, receptors on the cell surface are bound by an external ligand, and then subsequently brought inside the cell as a segment of the cell's membrane is bent inward and invaginates to "pinch off," becoming an internalized *vesicle* that contains that receptor (Figure 22.1). Once this occurs, the internalized receptor is usually neutralized because it can no longer bind to external ligands on the plasma membrane, and its ability to transmit biochemical signals is decreased from inside the cell.

While internalization, also known simply as *endocytosis*, is a very well-characterized phenomenon, its underpinnings have been decades in the making and began with the studies of the Russian scientist Ilya Metchnikoff in the late 1800s. Metchnikoff, who along with Paul Ehrlich received the Nobel Prize in 1908, pioneered studies on *phagocytosis* (Greek for "eating of the cell"). At the time there was great competition between a group of very well-established German pathologists, including the famed Rudolph Virchow and his followers, and Metchnikoff. The Germans were proponents of the involvement of parenchyma cells and primary lesions of the blood vessels as being responsible for inflammation, whereas these ideas were ultimately overturned by Metchnikoff, the "crazy" Russian outsider, who first demonstrated that starfish larva can engulf (phagocytose) rose thorns. Later, Metchnikoff posited that macrophage cells and other leukocytes serve as the first line of defense, by precisely such a mechanism—the engulfment of pathogens (bacteria, virions, and other foreign particles)

DOI: 10.1201/9781003202974-22

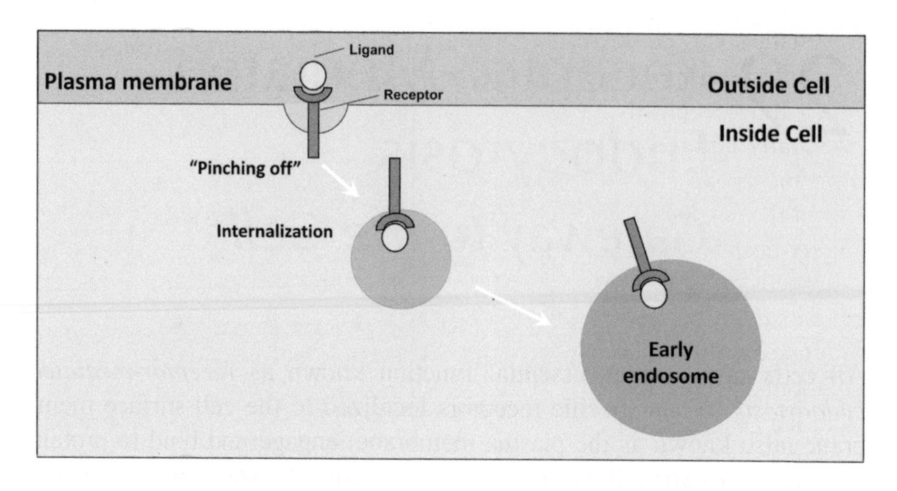

FIGURE 22.1 Receptor-mediated endocytosis. A ligand such as a growth hormone binds to a receptor expressed on the cell surface membrane, triggering an invagination process in which the receptor enters the cell within a spherical vesicle that has pinched off from the plasma membrane. The internalized vesicle later fuses with an intracellular compartment, known as the early or sorting endosome, from where it may be transported to the lysosomal pathway for degradation, or returned to the plasma membrane in a process known as *recycling*.

and their uptake into cells, where eventually they make their way to lysosomes and are usually neutralized and degraded (Silverstein 2011). Notwithstanding the huge implications in the immune response and in the process of inflammation, with macrophages playing a major role, phagocytosis and the so-called phagocytic theory were the first demonstration of a more general process by which cells are able to surround external material with their plasma membrane and internalize these membrane-enclosed structures into the cell. Indeed, demonstrating the existence of phagocytosis laid a cornerstone that subsequently transposed into one of the most important and studied events in the life of a cell—endocytosis.

As with many of the stories of the great discoveries in the modern history of biomedical science, discovering receptor-mediated endocytosis—or at least uncovering the details and mechanisms of this crucial process—took a metaphoric scientific village. Moreover, it also took continuous advances in technology to really move the field forward in a substantial manner. From Metchnikoff and his discovery of phagocytosis, some 50 years went by until Warren Lewis of the Carnegie Institution of Washington made a number of truly remarkable movies of live cells. Lewis, who was decades ahead of his time and undoubtedly would have delighted in observing the revolution in live cell imaging that has occurred over the last two decades in the wake of the discovery of the green fluorescent protein (see Chapter 24), coined the term "pinocytosis" (Greek for drinking by the cell) and observed that "Fluid entrapped by the ruffled pseudopodia enters the cell as globules which move centrally." In this succinct but accurate description, Lewis essentially laid the groundwork for the concept of vesicles that pinch off from the plasma membrane of the cell and fuse with larger endosomes within the cell interior. A clip from Lewis' work, which is of remarkable quality despite being from 1936, shows mouse macrophages in culture in the process of pinocytosis, or in the drinking of the surrounding fluid from the outside of the cell, and can be seen at the following link: https://www.youtube.com/watch?v=7qUCTgJ8C8s (Lewis 1936).

About a decade before the first reports were made by Roth and Porter of what later became known as clathrin-coated pits (Roth and Porter 1964), George Palade identified another specialized site on the plasma membrane that he called "plasmalemmal vesicles" (Palade 1953b). These structures, whose role in the cell remains somewhat controversial even today, are now known as *caveolae*, or small caves, based on their cavern-like appearance in electron micrographs, but at the time little was known about their role in the cell.

The study by Thomas Roth and Keith Porter published in 1964 in the *Journal of Cell Biology* was a true landmark paper that took advantage of electron microscopy to show that yolk protein in mosquito oocytes was taken up into cells "in a process akin to micropinocytosis," likely through "pit-like depressions on the oocyte surface" that are formed by invagination of the oocyte cell membrane (Roth and Porter 1964). Essentially, these findings, as described in the abstract from this paper, summarize the process of clathrin-dependent endocytosis. Indeed, at the end of the introduction to this paper, Roth and Porter note that the intention of the study was "to describe these pits and to present reasons for interpreting them as surface differentiations designed specifically for protein uptake and transport into the oocyte" (Roth and Porter 1964).

In this classic paper, the authors were able to demonstrate very clearly that during active yolk deposition, oocytes display 15 times more "pits" than usual, strongly suggesting that these structures are used to take up the protein into the cell. Moreover, the beautiful electron microscopy images capture a range of different stages for this uptake, showing the initial "depressions" of the pits at the plasma membrane, their invagination into vesicles within the cell, and even the various states at which one can easily envision that the pits are cleaved off from the plasma membrane to form independent, fully enclosed internal vesicles. By comparing coated structures that were still attached to the plasma membrane with those at varying stages of internalization, to those that separated from the membrane to become intracellular vesicles, Roth and Porter were able to come to the well-founded conclusion that these stages were all related, and were a continuum of the process that was initiated at the plasma membrane. Indeed, they realized that the various steps they observed were an indication of the dynamic process by which coated pits were internalized and subsequently released into the cytoplasm. Remarkably, these observations which were made from a series of purely static images now form the basis of our modern understanding of the dynamic way that receptors are internalized from the plasma membrane into cells.

Just over a decade later, building on the beautiful blocks laid by Roth and Porter, another group led by research partners Michael Brown and Joseph Goldstein would go on to receive the Nobel Prize in 1985 for their studies on the internalization of receptor-bound low-density lipoprotein (LDL) through the very same pits described by Roth and Porter (Anderson, Brown, and Goldstein 1977). In a very sophisticated study that used electron microscopy coupled with the uptake of LDL labeled by radioactive material, Anderson, Brown, and Goldstein were able to synchronize the uptake of LDL into the cell by using 4°C for binding to surface receptors, so that little or no internalization of the receptors occurred at this temperature. They then followed the 4°C binding by shifting the temperature of the cells rapidly back to 37°C to allow the synchronized uptake of the LDL-receptor complexes from the plasma membrane into the cell (Figure 22.2). The radioactivity was used to pinpoint where the LDL localized within the cell when examined by electron microscopy. These studies demonstrated that the coated pits are specialized regions of the plasma membrane that

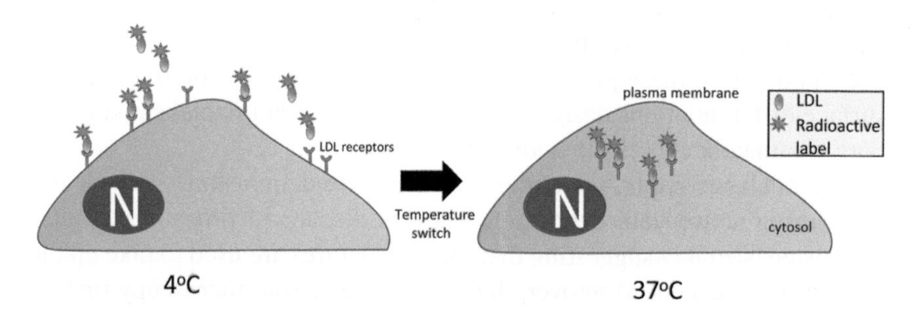

FIGURE 22.2 Schematic diagram illustrating synchronized internalization of low-density lipoprotein receptor. Low-density lipoprotein (LDL) that is tagged with a radioactive label (depicted by purple star) binds to LDL receptors at the cell surface at 4°C. Due to the cold temperature, the receptors remain on the cell surface and are not yet internalized into the cell. Once binding has occurred, excess (unbound) LDL is washed away and the cells are shifted to 37°C (temperature switch). This allows synchronized uptake of all the LDL receptors bound to the radiolabeled LDL into the cell interior.

are designed to take up internalized receptors by invagination and release them into the cytoplasm. Additionally, the study found that the coated vesicles undergo uncoating after a short time within the cytoplasm and these "newly-naked" vesicles fuse very rapidly with lysosomes.

In a very scientific and collegial manner, the authors gave credit to the decade-old study of Roth and Porter, and to Carpenter and Cohen who published similar results describing the epidermal growth factor receptor (EGFR) and its internalization via coated pits (Carpenter and Cohen 1976), hypothesizing that the very same pits might be responsible simultaneously for the internalization of both types of receptors.

Even 50 years after Roth and Porter published their findings, it remains difficult to overestimate the collective significance of these studies describing the process of receptor-mediated endocytosis. Brown and Goldstein's work on LDL and its receptor has led to a dramatically enhanced understanding of atherosclerosis and the mode by which cholesterol is taken up into cells and removed from the serum. Indeed, specific mutations in the receptor that impede its uptake (along with the uptake of the bound LDL and cholesterol) were determined to be a cause of some forms of familial hypercholesteremia (Goldstein 1989). Another example of the incredible significance of the findings of Roth and Porter comes from knowledge of the subcellular itinerary of the EGFR subsequent to its internalization via coated pits (Carpenter and Cohen 1976). In particular, EGFR has been described as a driver of oncogenic events in breast cancer, lung cancer, and glioblastomas (Sigismund, Avanzato, and Lanzetti 2018). The EGFR, which is a receptor that possesses intrinsic kinase activity and the ability to catalyze the covalent attachment of phosphate moieties to select tyrosine residues within its own amino acid sequence, can thus trigger a cascade of cellular events once it is "activated" through the binding of epidermal growth factor to the receptor at the cell surface. The phosphorylation event that is induced by receptor-ligand binding rapidly leads to intracellular signals being transmitted through the cytoplasm and on to the cell nucleus, where genes involved in cell proliferation, and/or differentiation and survival are turned on (Lemmon and Schlessinger 2010). The significance for cancer is undeniable. For example, the most common form of glioblastoma contains an EGFR variant known as EGFR type VIII, which lacks 267 amino acids in its extracellular domain and is incapable of binding epidermal growth factor but nonetheless remains constitutively active at the plasma membrane (Gan, Kaye, and Luwor 2009), thus highlighting the significance of endocytosis (or failure to carry out endocytosis) in this disease. While clinical trials to repair mutant EGFR in cancer patients have yet to be established, the use of CRISPR/Cas9 gene editing to test the potential of genetically manipulating the receptor as a proof-of-principle (Tang and Shrager 2016) has again demonstrated that the coupling of advances from basic science (such as CRISPR/Cas9 gene editing and knowledge from mechanisms of endocytosis and the functioning of EGFR) will likely elicit new strategies for the treatment of debilitating illnesses in the coming years.

23 Mitochondria
*The Metabolic Powerhouse of the Cell****

As will be revealed in the next and final chapter of this book on the green fluorescent protein (GFP), not all scientific discoveries have led to the generation of therapeutics—although it is anticipated that in many cases, over time clinical advances will eventually follow the basic discoveries. The discovery of mitochondria, an organelle whose existence was first observed over 150 years ago, is likely such an example.

In 1790, the French chemist Antoine Lavoisier wrote that organisms need to burn fuel in a manner similar to that of lamps and candles, noting

> In respiration, as in combustion, it is the atmospheric air which furnished oxygen ... but since in respiration it is ... the blood, which furnishes the combustion matter, if animals did not regularly replace by means of food ... that which they lose by respiration, the lamp would soon lack oil, and the animal would perish as a lamp is extinguished when it lacks nourishment.

Thus he can be depicted as the "father of metabolism" (Underwood 1943). It would take nearly 100 years before it was demonstrated by Eduard Pfluger, a German physiologist, that respiration does in fact occur within cells; Pfluger showed that respiration takes place outside the bloodstream in peripheral tissues (Culotta 1970), although his findings preceded the discovery of mitochondria and their involvement in the process. However, it was not until the mid-1850s that the structures that likely corresponded to mitochondria were first observed and documented, and in 1890 recognized by Richard Altmann as *bioblasts*, cytoplasmic structures within the cell that resembled bacteria and that "functioned as elementary organisms" (Ernster and Schatz 1981). It is of interest that this notion of mitochondria as a symbiotic organism living within eukaryotic cells was virtually ignored for nearly a century until Lynn Margulis (then Lynn Sagan) published a paper maintaining that mitochondria (as well as plastids and the eukaryotic flagellum) originated through an endosymbiotic mechanism in which prokaryotes or bacteria had been *ingested* by eukaryotic cells billions of

DOI: 10.1201/9781003202974-23

years ago (Sagan 1967). In other words, mitochondria initially thrived in host cells as independent, symbiotic organisms (where the mitochondria and cells each benefited from the other). Today these notions of eukaryotic organelles originating from endosymbiosis are largely accepted, if still poorly understood (Gray 2017).

The term *mitochondria*, in Greek meaning "threads" (*mitos*) and "granules" (*chondros*), was coined by German scientist Carl Benda in 1898 and described the appearance of mitochondria during spermatogenesis using a staining dye called crystal violet (Ernster and Schatz 1981). Indeed, these structures were subsequently observed in living cells by Leonor Michaelis using a dye known as Janus Green B, thereby demonstrating that the stained organelles observed existed in living cells and were not a post-fixation artifact (Ernster and Schatz 1981). The demonstration by Pfluger that respiration does not occur in the blood led to years of searching for a *respiratory pigment* postulated by Charles MacMunn in 1884, something akin to the hemoglobin that carries oxygen in erythrocytes (red blood cells), until David Keilin and his colleagues at Cambridge University in the United Kingdom identified the cytochromes. These are essential mitochondrial proteins that contain a heme (iron-binding) group, and are capable of carrying out electron transfer reactions from one molecule to another by oxidation (loss of electrons) or reduction (gain of electrons) on their heme iron (Keilin 1925, Keilin and Slater 1953, Keilin and Hartree 1945). The work of Keilin helped to refine the ideas of Otto Warburg (Nobel Prize awarded in 1931) at the Kaiser Wilhelm Institute for Cell Physiology in Berlin, Germany, who showed that respiration is related to particles found in extracts of liver derived from guinea pigs (Kohler 1973, Warburg 1913). Both Warburg and fellow German scientist Heinrich Wieland (Nobel Prize 1927) held strong ideas about the particulate nature of respiration, with Warburg favoring the involvement of a *respiratory enzyme* (he called it "*Atmungsferment*") involving the transfer of oxygen, and Wieland proposing the transfer of hydrogen (Ernster and Schatz 1981). Keilin's studies were crucial to understanding the function of mitochondria and his vision of a *respiratory chain* that involved mitochondrial dehydrogenase enzymes in the process of aerobic metabolism. Only once Keilin and his lab identified the *Atmungsferment* as a member of the cytochrome family did the key components of this respiratory chain begin to clarify (Keilin 1925).

Aerobic metabolism and the contribution of the mitochondria to respiration were still poorly understood. The major cellular source of energy, adenosine triphosphate (ATP), was not discovered until 1929 (Lohmann 1929), and the citric acid cycle (also known as the Krebs cycle) was discovered in 1937 by Hans Krebs (who was awarded the Nobel Prize in 1953) and his studies helped demonstrate how a key molecule known as acetyl coenzyme A leads to the production of two carbon dioxide molecules during the final stages of glycolysis (Krebs and Johnson 1937). However, these discoveries were made independently of mitochondria and the relationship of this important organelle to these biochemical pathways remained unknown.

As discussed earlier in the chapter detailing the discovery of the lysosome (Chapter 20), the advent of differential centrifugation to separate components (organelles) of the cell to study their contents and function led to dramatic new discoveries in the 1940s and 1950s. Indeed, using these methods, Albert Claude and the members of his laboratory were able to show that key enzymes localize to the mitochondrial fractions (Hogeboom, Claude, and Hotch-Kiss 1946). This was followed by reports that the Krebs citric acid cycle and the process of fatty acid oxidation occur in mitochondria (Kennedy and Lehninger 1949), firmly building the case for mitochondria as central organelles in the generation of energy for the cell.

In parallel with the advances that occurred in the study of other intracellular organelles, including lysosomes and the Golgi, a comprehensive understanding of mitochondria similarly benefited from the application of electron microscopy. Indeed, electron microscopy images determined that there is a double membrane surrounding the mitochondria, and demonstrated the existence of the *cristae*, the folds within the inner mitochondrial membrane (Figure 23.1; see schematic diagram and fluorescence micrograph) (Palade 1953a, Sjostrand 1953).

Together with the great strides in understanding the morphology and structure of this organelle, the next ~30 years cemented the role of mitochondria as the energy-generating center of the cell. In the early 1960s, a radical hypothesis was proposed by Peter Mitchell who postulated that through an electron transport chain there is a release of energy that helps generate a gradient of hydrogen ions (protons) across the mitochondrial membrane, leading to the release of additional energy that is captured and stored in ATP molecules (Mitchell 1961).

Beyond the crucial function of mitochondria in providing energy for the cell, over the last 30–40 years mitochondria have been significantly implicated in other vital cellular processes, in particular in the control of programmed cell death, also known as apoptosis. Indeed, it was discovered that cytochrome c, one of the cytochrome proteins postulated and identified by MacMunn and Keilin over 95 years ago, is released from mitochondria when apoptosis is induced in a cell, and it plays a role in triggering the apoptotic pathways (Liu et al. 1996, Li et al. 1997). Given that the failure of cells to undergo programmed cell death has been linked to excessive proliferation, this connects mitochondrial function directly to a host of oncogenic diseases.

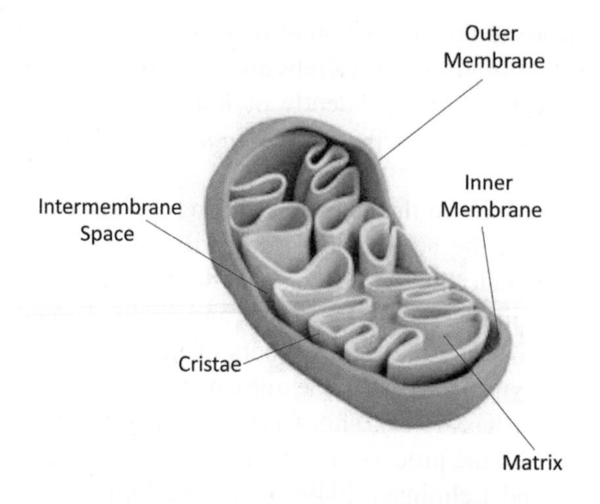

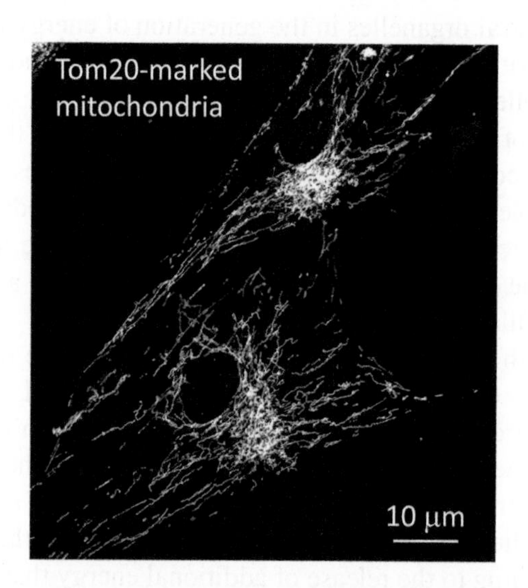

FIGURE 23.1 The mitochondrion. Schematic illustration of the mitochondrion, showing the outer mitochondrial membrane, intermembrane space, inner mitochondrial membrane, cristae, and mitochondrial matrix. Below, a fluorescence microscopy image of mitochondria in the cell. Human HeLa cells on glass coverslips were fixed and immunostained with antibodies against the mitochondrial outer membrane protein, TOM20. Image provided courtesy of the author.

Mitochondria, however, are not linked via apoptosis exclusively to cancer, but additionally to a variety of other illnesses including heart disease and neurodegenerative disorders such as Alzheimer's disease and Parkinson's disease. As such, although the discovery of mitochondria has not yet led to effective treatments and cures for the many illnesses arising from their dysfunction, there are numerous trials underway to explore the idea of treating diseases in which mitochondrial function is impaired (Wang, Karamanlidis, and Tian 2016). Indeed, there are well over 1,200 clinical trials listed on the NIH website ClinicalTrials.gov, many of which have been completed. These trials range from the analysis of mitochondrial DNA content (mitochondria have their own DNA, although they require many proteins coded by nuclear DNA for their function) and its correlation with severity and ontology of lymphoblastic leukemia in children, to the use of a "mitochondrial cocktail" that includes supplements, metabolite intermediates, and anti-oxidants to attempt to treat autistic children with mitochondrial dysfunction, and to the extraction of mitochondria from a pediatric cardiac patient's own skeletal muscle (*autologous mitochondria*) for injection into the heart to improve mitochondrial and cardiac function.

The discovery of mitochondria and continued new discoveries surrounding mitochondrial function that go far beyond the critical significance of this organelle as the key cellular energy source is an important lesson in the history of biomedical science; the more deeply researchers understand the basic science, the more likely it is that new and crucial drug treatments will arise. As aptly noted by Robert Chambers, writing in a book edited by Edmund Cowdry in 1924:

> In the first place, it is quite obvious that the investigation of mitochondria will never achieve the usefulness which it deserves as an instrument for advance in biology and medicine until we know much more of their chemical constitution as the only accurate basis for interpretation of our findings. In other words, we must wait upon the slow development of direct, qualitative cellular chemistry.
>
> **(Chambers 1924)**

Thus, rather prophetically Chambers summed up the next ~100 years of mitochondrial research, and indeed, all of biomedical research.

24 The Light at the End of the Tunnel

*Discovery of the Green Fluorescent Protein**

Not all basic scientific discoveries are immediately "translatable" to specific diseases or medical cures, but sometimes there are discoveries that have so much potential that it is clear that they will ultimately have important benefits to human health. The discovery of the green fluorescent protein (GFP) is one such discovery, and although to date its many exciting uses have been relegated primarily to research (rather than to medicine), the discovery of GFP is included herein due to its incredible value and future potential.

Although fireflies and their remarkable bioluminescence had been observed and marveled at for many centuries, it was Charles Darwin who poetically wrote in his diary about the fluorescence in the sea:

> While sailing in these latitudes on one very dark night, the sea presented a wonderful and most beautiful spectacle. There was a fresh breeze, and every part of the surface, which during the day is seen as foam, now glowed with a pale light. The vessel drove before her bows two billows of liquid phosphorus, and in her wake she was followed by a milky train. As far as the eye reached, the crest of every wave was bright, and the sky above the horizon, from the reflected glare of these livid flames, was not so utterly obscure, as over the rest of the heavens.
>
> **(Darwin)**

Darwin further noted that jellyfish, which are the source of the GFP molecule that has been co-opted as a major research advance about 150 years later, had their own intrinsic fluorescence possibly as a result of "a disturbed electrical condition in the atmosphere." Indeed, in 2008 the Nobel Prize in Chemistry was awarded jointly to three scientists whose research has led to the widespread use of the GFP (and a growing number of newly

DOI: 10.1201/9781003202974-24

discovered and related fluorescent proteins) in thousands of laboratories across the globe, and whose discovery and use has revolutionized biomedical research and altered the pace of discovery.

What is GFP? It is a protein initially discovered by Nobel laureate Osamu Shimomura found in an organism known as the Pacific Northwest jellyfish *Aequorea victoria* in 1962 (Shimomura, Johnson, and Saiga 1962). The original paper was somewhat understated, detailing a very biochemical approach to the purification of this jellyfish protein, and the measurement of its luminescence in a fairly technical manner. Indeed, most of the subsequent discussion in the paper focused on the mode of purification of the protein and the mechanisms of its luminescence. Remarkably, absent was any discussion of the revolutionary nature of their findings, and the mighty impact that GFP would have on biomedical research within the lifetime of Dr. Shimomura.

The GFP protein is a barrel-shaped molecule whose molecular weight is about 27 kilodalton—a protein of small to average size. Studies have identified the chromophore in the protein that emits the fluorescence as being derived from three residues, amino acids 65–67, consisting of serine-tyrosine-glycine (Tsien 1998). Nobel laureate Roger Tsien, who not only worked out the mechanisms of GFP function but also developed a series of GFP variants that have significantly enhanced the utility of GFP, was instrumental in elucidating the identity and mechanism of the chromophore. He found that once the protein properly folded, a series of chemical reactions induce the formation of the chromophore. As a result, when the GFP protein is illuminated or excited with light, it will in turn emit light at a higher wavelength, usually in the green realm of the visible spectrum, with the precise excitation and emission wavelengths depending on whether one looks at the original wild-type GFP or some of its many variants that were later developed (see Figure 24.1; schematic illustration and example of a GFP-tagged protein in a fluorescence micrograph image).

One might very well ask why this is so important, given that there are so many fluorescent dyes that can be chemically attached to proteins, or that there are antibodies that can be used to detect cellular proteins and detect their localization within the cell. GFP has major advantages that make it irreplaceable as a scientific tool. First, given that GFP has been cloned and its DNA sequence is entirely known so that it can be used in genetic engineering, researchers have taken advantage of GFP by using it as a "tag" and to genetically couple it to other proteins. What this means is that a generic Protein-X can now be coupled to GFP, rendering it Protein-X-GFP, or GFP-Protein-X, depending on which end of Protein-X the GFP is appended to. On the assumption that GFP is relatively "inert" compared to the protein to which it is coupled (Protein-X in this example), then

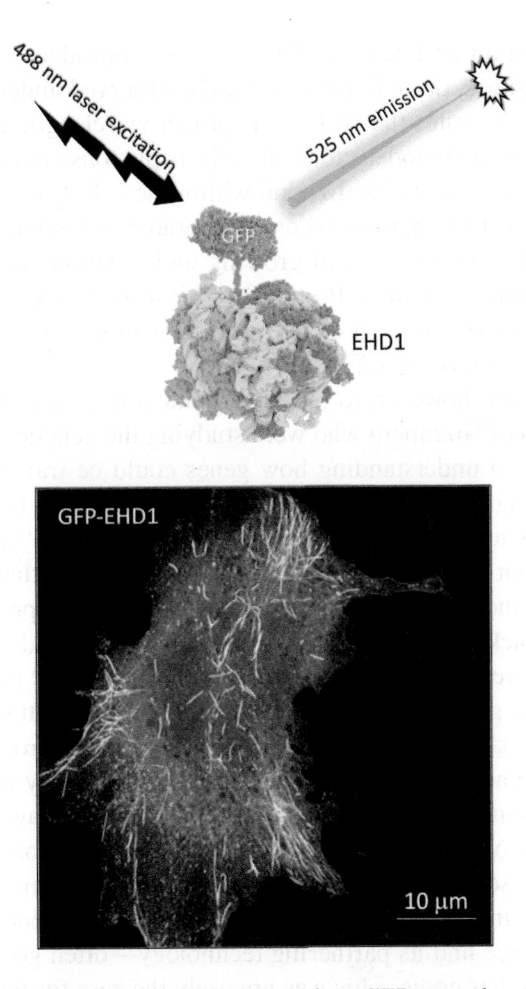

FIGURE 24.1 HeLa cells transfected with a GFP-tagged protein known as Eps15 Homology Domain protein-1 (EHD1). The schematic diagram illustrates how a 488 nm laser excites the GFP-tagged EHD1 protein, which subsequently emits light at a higher wavelength (~525 nm), which is detected by a fluorescence microscope. The lower panel shows a fluorescence micrograph obtained after HeLa cells were transfected with the GFP-EHD1 plasmid, and the green fluorescent-marked protein is imaged and observed along tubular recycling endosomes seen in white. Image provided courtesy of the author.

wherever Protein-X is normally found within the cell, the GFP can be visualized there, thus providing crucial information on protein localization and function. While this may not appear to be a substantial advance compared to simple microscopy using antibodies and dyes to detect protein localization in fixed cells, the most crucial advantage of the GFP is that a protein such as Protein-X-GFP can be visualized *in living cells, in real time*. Once the cDNA containing Protein-X-GFP has been introduced or transfected into cells in culture, those living cells can be observed under a microscope with a laser that excites the GFP at the proper wavelength. Accordingly, a microscope with a photodetector that collects photons emitted by the GFP can be used to record its localization within the cell. If the microscope is also equipped with a chamber whose temperature is kept at 37°C with 5% carbon dioxide (to keep the cell growing under optimal conditions), it is not only possible to localize Protein-X-GFP in real time, but to actually follow the movement of the protein over time within the cell by continuous excitation/emission of the GFP.

It is necessary, however, to take into context that in 1962, researchers such as Marshall Nirenberg who were studying the genetic code were still a long way from understanding how genes could be translated into proteins. Accordingly, it was difficult at the time to conceive how researchers might take advantage of Shimomura's discovery of this unique jellyfish protein. Without being able to identify the specific DNA that codes for the GFP protein, and without knowing how to induce its expression in cells, GFP simply lacked the remarkable utility that it has today. At the time researchers were limited to biochemically enriching and purifying small amounts of the protein to be studied, but the ability to attach its DNA to that of a mammalian protein and form a fusion protein product that could be monitored and followed either throughout its itinerary in cells or during the development of an organism remained decades away. This again highlights how over and over throughout the history of biomedical science it is possible to see that discoveries that are made prematurely—before the erection of the infrastructure that comes from a fundamental understanding of the science and its partnering technology—often go unappreciated until their time has come. This was precisely the case for the discovery of the GFP.

Three decades later, Prasher and his co-workers published a paper detailing the cloning of GFP (Prasher et al. 1992), opening the gateway for researchers to begin to take advantage of the remarkable properties of this "enlightening" protein, pun intended. The molecular cloning of the complementary DNA (cDNA) of GFP, which also furthered the biochemical understanding of how GFP forms its fluorophore and is able to emit light, was almost immediately followed by additional characterization by Inouye and Tsuji (Inouye and Tsuji 1994) and remarkably, the first demonstrations that it can be expressed in other organisms such as the worm, *C. elegans*, and used to visualize neurons in living animals by the group of Nobel laureate Martin Chalfie (Chalfie et al. 1994). Today in 2021, a quick search of biomedical papers using either GFP or green fluorescent protein as a keyword yields almost 60,000 published and indexed papers, which of course do not include all of the variants and the many other fluorescent proteins that have been identified and engineered as a direct result of research on GFP.

At a cellular level, GFP and fluorescent-tagged proteins have allowed researchers to mark specific organelles: for example, by coupling GFP to a protein exclusively localized to the mitochondrial membrane, scientists have found a unique way to study specific organelles, from the mitochondria (in this example) to the nucleus, Golgi complex, endoplasmic reticulum, lysosomes, endosomes, peroxisomes, and even the plasma membrane (Figure 24.1; see example of GFP-tagged protein in fluorescence micrograph). Researchers take advantage of the unique properties of GFP, its "photosensitivity" or tendency to "go dark" and become "bleached," losing its fluorescence when repeatedly excited by a laser beam. This feature of GFP proteins allows researchers to purposefully *photobleach* the GFP, causing it to lose its ability to emit fluorescence, thus facilitating kinetic measurements of protein movement from location to location. For example, if a GFP-tagged Golgi protein is fluorescing at the Golgi, and subsequently photobleached, then any recovery of fluorescence observed at the Golgi over time (when a living cell is imaged under the microscope at 37°C) must then reflect movement of GFP-tagged protein to the Golgi complex from somewhere else in the cell, since the photobleaching is considered to be, for the most part, irreversible. Using such GFP-based technologies, researchers can also measure the rates of GFP-tagged proteins that are secreted or degraded, in addition to other fates.

Beyond the incredible versatility of GFP at the cellular level, and beyond its remarkable facilitation of mapping gene expression in animals that was first illustrated by Chalfie and his co-workers, GFP-based technology has led to revolutionary new ways to study cancer and other diseases. For example, it is possible to use non-invasive imaging of cancer cells in mice or other animal models that express GFP-tagged proteins to closely monitor and determine the efficacy of cancer treatments in real time, and even examine the mechanisms by which these treatments work (Hoffman 2015). While these technologies are not yet in use for human cancer patients, one can easily envision the future advances that GFP technologies will likely bring.

The potential for GFP and its variants (yellow fluorescent protein, cyano fluorescent protein, red fluorescent protein, and others) in biomedical research is undeniable. As noted by Chalfie, "Scientific inquiry starts with observation. The more one can see, the more one can investigate. Indeed, we often mark our progress in science by improvements in imaging" (Chalfie 2009). He then went on to list the various researchers and Nobel Prizes awarded for imaging techniques, including X-rays and others. Indeed, it would be remiss to ignore the more recent 2014 Nobel Prize in Chemistry awarded to Eric Betzig, Stefan Hell, and William Moerner, whose studies have allowed researchers—using GFP technologies in some cases—to "break the diffraction barrier" and obtain resolution beyond what was considered possible by light microscopy. Moreover, as with many of the great biomedical discoveries throughout history, it began with observations of luminescence centuries ago, progressed to the purification and isolation of the GFP protein about 70 years ago, and as technology developed and the genetic code was solved, moved into a new era of harnessing GFP for a variety of uses including single molecule super-resolution imaging. Observing, identifying, purifying, studying mechanisms, cloning, expressing in different cells and organisms, improving and making brighter, more stable, more useful variants—all of these took a village of researchers. And undoubtedly, an even larger village will continue to take advantage and find exciting new and even more "illuminating" ways to use GFP for medical purposes in the years and decades ahead.

Conclusions

While this book cannot capture all of the wonderful discoveries in biomedical research over the last 150 years, the hope is that with the select examples highlighted within, it will be possible to attain a glimpse of the many ways in which the scientific enterprise is advanced. Each individual discovery is a world of its own, and yet there are recurring themes in science that have led to a significant number of these key findings. Many of these discoveries have been translated over the years into an understanding of pathways that have led to the development of new drugs and treatments for numerous illnesses, although it should be noted that some discoveries have yet to provide clear breakthroughs in the realm of clinical medicine.

One important lesson from science is that once knowledge has been gained, scientists and the general public rapidly lose interest in *how* that knowledge was attained. Most people are only interested in how the knowledge may be applied. Indeed, in his article about the career of Theodor Boveri, Florian Maderspacher notes that philosopher Ludwig Wittgenstein wrote that his reader should reach a point where "he must so to speak throw away the ladder, after he has climbed up on it" (Maderspacher 2008). With all due respect to the brilliance of Wittgenstein, in the realm of biomedical science, this misses the point entirely. As this book strives to emphasize, there is much to be learned from the way that scientists have achieved their scientific advances; understanding the scientific thought process, the way in which scientists have tackled problems in the past, is key to training the next generation of scientists. While the technologies continually advance and what researchers today deal with will differ from what they dealt with 100, 50, or even 10 years ago, the *mode* of problem solving and critical thinking remains the same. For this reason, scientists *should not* kick away the ladder upon climbing higher, but rather haul it along so that perhaps it can be used to attain new scientific heights in the next great challenge.

Another important lesson can be derived from Nobel laureate Aaron Ciechanover (2004) who was instrumental in working out the ubiquitin degradation pathway (Chapter 21). Ciechanover noted in his Nobel speech that in the course of his training,

> I also learnt to become a long books author rather than a short story writer: I learnt not to be opportunistic but rather to adhere to a project,

> to dig deeply into a problem, to resolve it mechanistically, to unravel complex mazes – peeling them like an onion, not to be tempted to be dragged after fashions
>
> **(Ciechanover, 2005)**

The notion that many people hold that scientific discoveries are made suddenly, on the spur of the moment, by one big *Eureka* idea is not borne out by a study of the history of science. Far more commonly, as noted by Ciechanover, it was a dedicated and consistent effort to address a given problem. In many cases, the more research that was carried out on a particular problem by different groups of researchers, often competing and spurring each other on, the more likely the chance for generating new knowledge and advancement. In other cases, as the examples in this book have strived to illustrate, technology sometimes served as a barrier for scientific advances. But perhaps the most important idea is that scientific discoveries have typically come from findings and observations that were made and advanced by a *village of scientists* over years, decades, and even centuries.

There are frequent disagreements within the scientific community about the best ways to support good science. The US grant system run by the National Institutes of Health is the *gold standard* for scientific research in the US, whether that science is basic research or more translationally/clinically oriented. To obtain support from this governmental agency, researchers are typically required to submit highly detailed research proposals that are known as *grant applications*. The preference is usually for research plans that are *hypothesis-driven* (meaning that there is a preconceived idea of what the potential outcome is likely to be) as opposed to "*discovery-based*" (meaning that the plans are more open-ended and screen a wide range of potential outcomes with less pre-conception). Both methods of scientific research are valid, and can be used in parallel to advance science, depending on the type of questions being asked. The key, as always, is in the rigor of the science. While hypothesis-driven studies are often advantageous because they necessarily force researchers to become skilled in planning projects and forecasting potential caveats and their solutions in advance, some institutes at the National Institutes of Health are coming to the notion that researchers may also benefit from having the freedom to pursue serendipitous studies—as they so frequently in the history of science have led to significant discoveries.

As a final note, just as valuable as the understanding of how conclusions regarding basic discoveries are made is the idea that from the time a discovery is made until it can be translated into medicines and treatments, it often takes many years. For this reason, it is essential that today's governments support today's basic scientists—so tomorrow's cures will be in the pipeline.

Glossary of Terms

A

***Adoptive cell transfer (ACT)*:** An immunological therapy where T lymphocyte cells derived from a patient's blood are grown and expanded and sometimes genetically manipulated in culture prior to their return to the patient to bolster the immune response against the disease (usually cancer).

***Adenosine triphosphate (ATP)*:** An organic molecule in the cell considered to be a primary source of energy, which is released upon hydrolysis of the molecule to adenosine diphosphate (ADP) and adenosine monophosphate (AMP).

***Amino acid*:** One of the 20 organic compounds made of primarily carbon, hydrogen, oxygen, and nitrogen that make up the building blocks of proteins. All amino acids have amino terminals (with amine chemical groups) and carboxyl terminals (with carbon-oxygen groups) that link to one another to form peptide bonds and proteins, and each amino acid is distinguished by unique side chains of different chemical composition.

***Anaplastic large T cell lymphoma (ALCL)*:** A relatively rare type of non-Hodgkin's lymphoma (cancer of the white blood cells) that occurs in T lymphocyte cells.

***Angiogenesis*:** The generation of new blood vessels. Angiogenesis was once thought to occur exclusively in developing organisms (and in women's reproductive cycles) but is also crucial for the growth of solid tumors.

***Antibody*:** Also known as an immunoglobulin, an antibody is a protein that is generated in B lymphocyte cells that differentiate into plasma cells. It is secreted into the serum and has the capability to bind and neutralize select antigens (toxins or foreign invaders). Antibodies are unique and diverse, and highly specific, and form the backbone of the body's immune response.

***Antibody-dependent cell cytotoxicity (ADCC)*:** A mechanism by which the immune system eliminates target cells or foreign invaders after they have been bound or marked by specific antibodies. Once the variable recognition region of the antibody binds its target, the antibody's invariable Fc region is coupled to effector immune cells that lyse the target.

***Antigen*:** An immunological term for a molecule (or region of a molecule) of a pathogen that elicits an immune response when recognized by antibodies or cells of the immune system.

***Antisense strand*:** The DNA double helix is comprised of two strands of DNA: the coding strand (known as the sense strand) that comprises the genes that are transcribed and translated into proteins, and the antisense strand which is used as a template to copy the DNA during the process of replication.

***Apoptosis*:** A type of induced programmed cell death that leads to the release of enzymes known as caspases from the mitochondria, culminating ultimately in a variety of cellular effects that occur prior to death, including nuclear fragmentation and membrane blebbing.

***Autoradiography*:** A technique in which X-ray film is used to detect radioactive particles and thus reflect the distribution or localization of a radioactively labeled substance, such as a protein, nucleic acid, or sugar compound.

B

***B cell*:** A key lymphocyte or white blood cell that expresses antigen-specific immunoglobulins and is a precursor of the plasma cell that secretes antibodies.

***Biosynthetic transport*:** The transport pathway of newly synthesized proteins from the Golgi complex to endosomes and lysosomes where they reside and function.

***Blastocyst*:** A stage of mammalian embryonic development with about 150–300 cells that includes the inner cell mass that forms the embryo and the trophoblasts that comprise the placenta.

C

***Candidate approach*:** A scientific approach in which the several key candidates are evaluated, as opposed to an "unbiased approach" where a wider net is cast to screen or test a greater number of possibilities.

***Cas9*:** This stands for CRISPR associated protein 9, an endonuclease enzyme that is found in bacteria that serves as part of the bacterial adaptive immune system involved in cleaving the DNA of invading viruses. This protein has been co-opted by researchers for use in genetic engineering in mammalian cells, a technique known as clustered regularly interspaced short palindromic repeats (CRISPR).

***Cell fractionation*:** This is a technique used by researchers in which the centrifugation or high-speed spinning of cell lysates leads to separation of cellular components or organelles into fractions based on biophysical properties, allowing localization of proteins and other cellular macromolecules.

***Cell-free system*:** This is an *in vitro* experimental system that relies on purified components and molecules to test a biological question without using whole cells. It is often advantageous because it is "reductionist" methodology and typically shows less variability than when using cells or whole organisms. Its primary disadvantage is that it might not reflect the true complexity of events within a whole cell.

***Cell theory*:** The notion that cells are the primary building blocks of tissues and that individual cells are comprised of three major components: a nucleus, fluid, and a surrounding wall or membrane.

***Central dogma*:** The central dogma of biology as outlined by James Watson maintains that DNA self-replicates, and it codes for protein by first being transcribed to messenger RNA (mRNA) which is then translated into a chain of amino acids (a protein).

***Central flagellum*:** The central flagellum was one of the initial terms used to describe the primary cilium, although the primary cilium differs from other cilia in the cell in that it is not motile like flagella.

***Centrifugation*:** This is a technique for the separation of molecules (or cells) by density, size, shape, or viscosity by spinning at high velocity.

***Centrioles*:** A pair of barrel-shaped proteinaceous (and nearly identical) organelles normally localized near the nucleus of eukaryotic cells that make up a crucial part of the centrosome and play a key role in cell division and organization of the cellular microtubule network. The older centriole, known as the mother centriole, is the starting point for biogenesis of the primary cilium.

***Centrosome*:** A membrane-less organelle typically localized near the nucleus comprised of two proteinaceous centrioles known as the mother centriole and the daughter centriole, as well as a surrounding pericentriolar matrix comprised of many proteins. It plays a key role in cell division and organization of the cellular microtubule network.

***Chimeric antigen receptor T (CAR-T)*:** A treatment based on genetic engineering done on a patient's T lymphocytes, which are removed from the blood and isolated. The genetic engineering is done so that the T cell antigen receptor on the surface of the cells is modified to selectively attack a specific target, such as a protein that is highly expressed on the surface of cancer cells.

***Chromosome*:** A long segment of DNA in the nucleus of the cell that contains genetic material or genes. Human cells typically have 23 separate chromosomes.

***Ciliogenesis*:** The process of generating a primary cilium in a cell, usually induced by nutrient deprivation or serum starvation. The primary cilium extends from the mother centriole as a microtubule-based structure that is surrounded by a membrane derived from the fusion of vesicles that forms a larger vesicle known as the ciliary vesicle.

***Ciliopathy*:** A genetic disease related to the primary cilium, often resulting in developmental disorders. Hundreds of genes have been established as ciliopathy-related genes.

***Citric acid cycle (also known as the Krebs cycle or tricarboxylic acid cycle)*:** A series of reactions that occurs within the mitochondrial matrix in which acetyl coenzyme A is consumed and leads to nicotinamide adenine dinucleotide (NAD) being reduced from NAD^+ to NADH, which is ultimately used to generate ATP in the electron transport pathway that can be stored or used as energy.

***Clathrin-coated pits*:** Depressions or inwardly curved pits in the cell membrane that are coated with a protein known as clathrin which forms unique "cages" for the internalization of receptors from the cell surface. This is considered to be a major pathway in which receptors are internalized from the surface membrane of the cell, once the clathrin-coated pits have been severed from the plasma membrane and internalized in vesicles within the cell.

***Clonal selection theory*:** An immunological theory that B lymphocyte cells maintain receptors for the diverse array of potential antigens or pathogens, and that once a given cell that expresses a specific receptor on its cell surface recognizes a given antigen, that clone of B cells will proliferate and expand to meet the antigen threat.

***Cloning*:** While "cloning" means to make a copy and is often used in the context of a whole organism (i.e., Dolly the Sheep), most researchers usually refer to cloning as a technique in genetic engineering where a gene or segment of DNA is generated, copied, amplified, and/or inserted into a plasmid or vector for the purpose of expressing it as a protein in cells.

***Clustered regularly interspaced short palindromic repeats (CRISPR)*:** While this is essentially a bacterial defense system against invading viruses, it usually now refers to its being co-opted as an extremely versatile method for genetic engineering in eukaryotic cells that allows precise cutting of genomic DNA for the purpose of modifications.

***Codon*:** A group of three consecutive nucleotides that code for a single amino acid. Since the genetic code has 4 nucleotides and there are 3 bases per codon, there are $4 \times 4 \times 4 = 64$ possible codons, with 61 combinations coding for the 20 amino acids, and 3 coding for a stop codon to terminate translation.

***Complementary DNA (cDNA)*:** This is a DNA copy of a messenger RNA (mRNA) sequence that has been "reverse transcribed" back into DNA by a DNA polymerase enzyme. It is generated by certain RNA viruses as part of their life cycle, but frequently used by molecular biologists to express proteins in cells.

***Controls*:** As part of an experiment, controls are included to minimize the possibility that factors other than those being tested might affect the results and conclusions of the experiment.

***Cristae*:** The folds within the inner mitochondrial membrane.

***Cytochrome*:** Belongs to a family of important mitochondrial proteins that contain a heme group and can catalyze the reduction or oxidation of their heme iron to perform electron transfer reactions.

***Cytology*:** The study of the cell (rather than at the level of the whole organism).

***Cytoplasm*:** The aqueous environment inside the cell, composed of water, salts, and organic molecules. Intracellular organelles such as the nucleus, lysosomes, endosomes, Golgi complex, and mitochondria have limiting membrane that separates them from the cytoplasm.

D

***Degradation*:** Usually refers to proteins and their cleavage into small peptides or individual amino acids, either within the lysosomal lumen or upon delivery to the proteasome.

***Degron*:** A short sequence motif of a protein that confers degradation of that protein, typically through the ubiquitin pathway.

***Differentiation*:** A process in development in which a less specialized cell undergoes maturation and becomes more specialized in its function. For example, a stem cell is an undifferentiated cell and, once it commits in differentiation, can become a muscle cell, kidney cell, retinal cell, etc.

***Discontinuity*:** An evolutionary term for variation describing individuals or traits that fall into discrete classes, as opposed to continuity where there is a complete range of change across an entire spectrum with no distinct classes.

***DNA*:** Also known as deoxyribonucleic acid, it forms a double helix and is comprised of the four nucleotides adenosine, thymidine, cytosine, and guanine, and molecules of DNA serve as the primary genetic material in living organisms.

***DNA polymerase*:** This is the enzyme that copies the DNA when the cell needs to undergo division (mitosis) and the DNA must be replicated so that the daughter cell can receive a complete copy.

***DNA replication*:** The process of copying the DNA (carried out by DNA polymerase) prior to cell division.

***Dominant*:** In most organisms there are two alleles, each of which contains genetic material passed from the parents. Dominant refers to a gene that will be passed on to offspring even if the other allele differs.

E

***Electron transfer*:** The transfer of electrons from a molecule, atom, or ion to another molecule. This includes both oxidation and reduction, known as redox.

***Electrophoresis*:** A technique for the separation of proteins, DNA, or RNA (or other macromolecules) based on their charge and size through a gel matrix.

***Embryonic stem cells*:** These cells are pluripotent in that they have the ability to give rise to cell types representative of all the tissues of the embryo and adult upon differentiation, and they continue to divide in a process of self-renewal. They are derived from the blastocyst inner cell mass of the embryo.

***Endoplasmic reticulum*:** This is a network of interconnected membranes in eukaryotic cells. It has a primary function in anchoring ribosomes that translate proteins destined for the endocytic and secretory pathways; such translated proteins go from the endoplasmic reticulum to the Golgi complex before being transported to their final destinations.

***Endosome*:** An organelle in the endocytic system surrounded by a lipid bilayer. Endosomes can be generally categorized as early/sorting endosomes that receive incoming vesicles and sort cargo receptors, as late endosomes in the lysosomal degradation pathway, or as recycling endosomes that are involved in the return of receptor cargo to the plasma membrane.

***Endosymbiosis*:** A process of symbiosis where two organisms benefit from a close physical relationship with one organism living inside another.

***Enzyme*:** A protein that catalyzes a biochemical reaction. In some instances RNA can also serve as an enzyme.

***Enzyme replacement therapy (ERT)*:** This is a method to replace a missing or dysfunctional enzyme, often one that resides within the lysosomal lumen of the cell, usually by injecting the purified enzyme into the bloodstream so that it is taken up into cells by receptor-mediated endocytosis. Once internalized, the vesicle carrying the enzyme fuses with endocytic compartments and eventually reaches the lysosome and functions in lieu of the missing enzyme.

***Epidermal growth factor receptor (EGFR)*:** An important cell receptor that binds to epidermal growth factor and transduces signals into the cell interior.

***Epitope*:** This is the portion of an antigen, a molecule recognized by an antibody, that is specifically recognized and bound by the antibody.

F

***Focus assay*:** Since cells that are transformed with oncogenes will continue to grow and not be inhibited when coming into contact with neighboring cells, when grown on a confluent monolayer of non-transformed cells they will form into dense and elevated foci that can be visualized and counted. This assay is used to determine whether cells are transformed.

G

***Gene*:** A hereditary unit comprised of DNA nucleotides that typically leads to the translation of a specific protein. It is part of a larger unit known as a chromosome, which contains multiple genes.

***Gene drive*:** The act of causing, by genetic engineering technologies such as CRISPR, a specific gene (or set of genes) to be propagated within a certain population.

***Genetic code*:** This is the code that dictates which codon (set of three nucleotides) specifies what amino acid will be inserted into a protein, and in what sequence. Given that there are four types of nucleotides and three nucleotides in a codon, there are $4 \times 4 \times 4 = 64$ combinations of codons. Sixty-one of these code for one of the 20 different amino acids, whereas as 3 codons signal termination of the protein translation process.

***Genetic engineering*:** This is a general term that refers to the manipulation of DNA, usually to insert, remove, modify, or otherwise alter expression of proteins in a cell or organism. The most recent and promising method for genetically altering cells and organisms is by CRISPR.

***Germline theory*:** A theory to explain the vast diversity of antibodies, maintaining that each antibody was encoded by a completely distinct gene in the germline. This theory was proven to be incorrect as the genome is not large enough to code for the necessary diversity in antibodies.

***Golgi complex*:** A membrane-bound organelle with an unusual morphology of a series of consecutive stacks. The Golgi complex is a key organelle for executing post-translational modifications on newly synthesized proteins, and it is crucial for the secretion of proteins from the cell, the delivery of receptors to the cell surface, and the targeting of proteins to organelles within the endocytic pathways (to endosomes and lysosomes).

***Green fluorescent protein (GFP)*:** This is a jellyfish protein from *Aequorea victoria* that is unusual in that when illuminated with a laser at the correct wavelength, it will emit visible light in the green wavelength. A common use for the GFP is to genetically fuse its cDNA to that of other proteins, so that its fluorescence can be used to localize the fusion protein. Its remarkable versatility allows the GFP to be used in living cells and living organisms to study protein movement and function.

***Guanine triphosphate (GTP)*:** This is a molecule that is rich in energy, and when hydrolyzed to GDP, the energy released catalyzes various reactions in the cell. In this sense, GTP functions similarly to ATP.

H

***Harvey ras (H-ras)*:** This gene codes for a GTPase known as HRAS, a protein that hydrolyzes GTP to GDP. This protein transmits proliferation signals to the cell when growth factors bind to their receptors on the cell surface. Originally identified as part of a murine leukemia virus as an oncogene, mutations in the endogenous human *H-ras* gene can lead to a variety of cancers.

***Hayflick limit*:** This is a limitation in the number of divisions that a non-immortalized cell in culture can undergo. The actual number depends on the cell type and conditions of growth but is usually 40–60 times. The reason for the existence of this limitation was only evident once telomeres had been discovered and their serial shortening with progressive divisions was observed.

***Helicobacter pylori*:** A bacteria that causes infections in the stomach leading to gastritis, peptic ulcers, and potentially stomach cancer. Its role in ulcers was not discovered until the early 1980s.

Hematopoietic stem cell: A stem cell with the capability of differentiating into any type of blood cell, including erythrocytes (red blood cells), and white blood cells such as T and B lymphocytes and leukocytes. Hematopoietic stem cells are typically localized to the bone marrow, where the process of hematopoiesis occurs.

Hemoglobin: A protein expressed in red blood cells that carries oxygen from the lungs to the tissues in the body and is involved in the retrieval of carbon dioxide from the tissues and its transport back to the lungs.

Hybridoma cell: A type of cell created by the fusion of two different cells together. Hybridomas are typically used to make monoclonal antibodies, where B cells from a mouse spleen are fused with myeloma cells that continually divide to create antibody-generating hybridoma cells that are immortal and continuously divide and secrete antibodies.

I

IgG: Also known as immunoglobulin G, this is the primary type of antibody isotype in the blood serum that is generated 7–10 days after an infection.

IgM: Also known as immunoglobulin M, this antibody isotype is the largest in size and also the first one generated after exposure to an antigen. Its presence is often used to diagnose an active infection, as within 7–10 days it undergoes "isotype switching" to IgG, so that over time the levels of IgM dwindle, while those of IgG are maintained.

Immortal clone: This refers to a clone of cells that possesses the ability to continuously replicate when cultured *in vitro*. Immortal clones are not subject to the Hayflick limit.

Immune surveillance: This term refers to the long-held view that the immune system continually monitors and detects malignant cells (or those on the cusp of becoming malignant) and targets them for destruction before they grow and take hold.

Immunoglobulin: An antibody molecule, either of the type IgM, IgG, IgE, IgA, or IgD.

In vitro: In a test tube or purified system without the presence of live cells (note: this term is sometimes also used to describe experiments with cells in culture, as opposed to a whole organism).

In vivo: In a living organism.

Induced pluripotent stem cell (iPS cell): A pluripotent stem cell that can be generated from a somatic cell (such as a fibroblast or other adult differentiated cell type) and then differentiated into any cell type.

Insulin: A protein that is secreted by pancreatic cells and regulates blood glucose levels (therefore known as a hormone). Failure to generate sufficient insulin (or cellular insensitivity to insulin) causes diabetes, or elevated blood sugar levels.

Internalization: A process by which receptors on the cell surface are bound by their specific ligand and consequently a portion of the membrane "invaginates" and is cut from the cell surface so that the receptor is now attached to a membrane-bound vesicle inside the cell cytoplasm.

Isotype switch: The switching from one immunoglobulin isotype to another. For example, antibodies are typically generated initially as IgM molecules and after 7–10 days undergo an isotype switch to IgG.

K

Kirsten Ras (K-ras): This gene codes for a GTPase, a protein that hydrolyzes GTP to GDP, known as KRAS. This protein transmits activation signals to the cell nucleus leading to proliferation. Originally identified as part of a murine leukemia virus as an oncogene, mutations in the human *K-ras* gene can lead to a variety of cancers.

Knock-down: Refers to the process of preventing the expression of a specific protein, usually by disrupting the transcription/translation of its gene. This is a method frequently used to study protein function.

Knock-out mice: The disruption of a gene for a specific mouse protein so that the animal lacks expression of that protein. This is a method frequently used to study protein function in a model organism, or to study disease caused by dysfunctional proteins in a living organism.

Koch's postulates: Based on the criteria set forth by bacteriologist Robert Koch, these postulates are criteria that determine whether a specific microorganism is the cause of a disease. They specify that: (1) the microorganism must be detected in individuals with the disease, but not healthy individuals, (2) the microorganism must be cultured from the individual bearing the disease, (3) it must be possible to transfer the disease to a healthy individual by inoculation with the cultured organism from the diseased individual, and (4) the organism must be cultured from the newly inoculated and infected individual and matched to the original organism.

Krebs cycle: Also known as the citric acid cycle. This is the pathway discovered by Hans Krebs who demonstrated how a key molecule known as acetyl coenzyme A leads to the production of two carbon dioxide molecules during the final stages of glycolysis.

L

***Lenhossek–Henneguy hypothesis*:** This hypothesis maintains that the primary cilium is derived from the centrosome and that therefore a ciliated cell cannot undergo cell division when its centrosome is engaged in ciliogenesis.

***Ligand*:** This is a molecule that binds to another molecule and is commonly used to describe secreted or cell surface proteins that specifically bind to the extracellular region of receptors on the surface of another cell.

***Low-density lipoprotein (LDL)*:** Particles comprised of protein and fats that come from the liver, including a phospholipid monolayer, triacylglycerol, and cholesterol, circulate in the bloodstream and deliver the cholesterol and lipids to tissues. It is internalized into cells through its binding to LDL receptors on the cell surface.

***Lysosomal hydrolases*:** Enzymes that reside within the lysosomal interior (lumen) whose function is the degradation of proteins and macromolecules that are transported to the lysosome.

***Lysosomal storage disorder*:** Inherited metabolic disease resulting from an accumulation of certain macromolecules (proteins, lipids, etc.) in the lysosomal lumen of the body's cells due to a deficiency or dysfunction of a specific enzyme.

***Lysosome*:** The major membrane-bound organelle in cells that plays an essential role in the degradation of proteins and other macromolecules. The lysosomal lumen contains specific enzymes that carry out the degradative functions of this organelle.

***Lysosome-related organelle*:** An organelle that carries out a dual role both in degradation of proteins and macromolecules (lysosomal function) as well as an additional specialized role. Examples of lysosome-related organelles are melanosomes (that carry pigment in skin cells and serve as lysosomes) and lytic granules (that secrete vesicles to target select cells and serve as lysosomes).

M

***Macrophage*:** A large specialized phagocytic cell differentiated from monocytes, the macrophage is a key cell of the innate immune response that engulfs or phagocytoses foreign particles and invaders, degrades them, and serves as an antigen presenting cell to activate T cells.

***Mannose-6-phosphate*:** A specific molecule that is post-translationally added to certain lysosomal enzymes as they traverse through the Golgi complex. The addition of the mannose-6-phosphate moiety is a signal that allows binding of the enzyme to the mannose-6-phosphate receptor, and it is subsequently packaged into vesicles and transported to the endocytic pathway from where it reaches the lysosome.

***Mendelian genetics*:** These are the genetic principles based on the discoveries of Gregor Mendel, that stipulate that genetic characteristics are inherited from each parent (known as alleles), and when they differ in the offspring (the offspring receives a different allele from each parent) then one allele is dominant over the other (the one that exhibits its characteristics). Mendel also demonstrated that the alleles are passed down by a random process known as random segregation, and that the traits or genes are unlinked to one another. He also maintained that characteristics that are passed down through this (genetic) inherited process are discrete rather than being a gradient (i.e., one color or another, but not a range of hues).

***Messenger RNA (mRNA)*:** The ribonucleic acid molecule that is transcribed from the DNA of genes that undergo expression and is decoded and translated into protein on the ribosome. The mRNA codons dictate the specific amino acids that are added to the nascent polypeptide chain.

***Mitochondria*:** Organelles separated from the cytoplasm in the cell by a double membrane. Mitochondria are key to biochemical reactions that produce energy, and are also crucial for programmed cell death, also known as apoptosis.

***Molecular disease*:** This expression was first used to describe sickle cell anemia, which was the first disease described where an aberrant protein was found to be causative.

***Monoclonal antibody*:** A highly selective antibody recognizing a single epitope. Monoclonal antibodies are usually generated in small rodents by injection of an antigen and fusion of the rodent spleen B lymphocytes with myeloma cells to create hybridomas that secrete antibody.

N

***N-end rule*:** Regulates the *in vivo* half-life of a protein by its N-terminal residue, with either acetylated or arginylated residues conferring targeting of the protein for degradation by the proteasomal pathway.

***Nuclein*:** The name used by Friedrich Miescher for the unique material that he enriched from the nuclei of cells which turned out to be deoxyribonucleic acid (DNA).

***Nucleotide*:** A molecule comprised of a sugar molecule (ribose for RNA or deoxyribose for DNA) that is attached to a phosphate group and a nitrogen-containing base. Nucleotides are the building blocks that link together to form RNA or DNA.

***Nucleus*:** A compartment inside the cell that has a surrounding membrane and normally contains the chromosomes and genes and genetic material of the cells, except during cell division.

O

***Oncogene*:** A gene (that may belong to a host cell or virus) that can transform a cell into a continuously replicating tumor cell.

***Oncogenesis*:** A process in which a cell (or organism) is transformed into a continuously dividing tumor cell.

***Organelle*:** A structure within the cell with a particular function and usually (but not necessarily) separated from the cytoplasm by a surrounding membrane bilayer. Examples include lysosomes, mitochondria, endosomes, the Golgi complex, etc.

***Over-expression*:** Transfection or introduction of cDNA coding for a specific protein into cells. This typically leads to higher-than-endogenous levels of expression of the protein, and thus over-expression of the protein.

***Oxidation*:** The loss of electrons of a molecule, atom, or ion. The molecule is then said to be in an "oxidized state."

P

***p53*:** A key gene that codes for a 53 kilodalton protein that regulates cell cycle and serves as a tumor suppressor gene, preventing oncogenesis when it is expressed.

***Penicillin*:** A family of bacteria-killing antibiotics derived from the mold *Penicillium.*

***Peptide fingerprinting*:** A technique by which proteins are purified, cleaved with proteolytic enzymes, and then the peptides of different lengths are separated and analyzed to determine their sequence.

***pH*:** A scale from 0 to 14 that measures the acidity (or basicity) of an aqueous solution; higher proton (hydrogen ion) concentrations mean a pH will be more acidic and have a lower pH value.

***Phagocytosis*:** A process in which a cell engulfs a particle (such as a bacterium, etc.) on the outside of the cell with its surface membrane. After engulfment, the surrounding cell membrane then detaches from the surface, thus internalizing the particle within a self-contained membrane-bound vesicle within the cell.

***Phosphorylation*:** A common post-translational modification of cellular proteins (and some lipids) that is carried out by enzymes known as *kinases* that covalently adds a phosphate moiety to either serine, threonine, or tyrosine residues. Phosphorylation often serves as a signal within the cell for turning on or shutting down a signaling pathway. Phosphorylation can be reversed by the action of phosphatases, which remove a phosphate group that has been added by a kinase.

***PI*:** This is the isoelectric point, the pH at which the net charge of molecule in solution is neutral.

***Plasma cell*:** A differentiated B lymphocyte that secretes antibody.

***Plasmid*:** A circular DNA molecule used to insert a cDNA representing a gene. Plasmids can be introduced into eukaryotic cells for their expression as proteins or be transformed into bacteria for amplification or generation of a protein.

***Polyclonal antibody*:** This type of antibody will recognize multiple epitopes (often of the same antigen molecule) because there are multiple different plasma cells secreting differing antibodies. Humans generate a polyclonal type of response when given a vaccination with an antigen or when encountering a pathogen.

***Polymerase chain reaction (PCR)*:** A technique for the amplification of DNA. This technique takes advantage of the characteristics of DNA and a unique DNA polymerase from the thermophilic bacteria *Thermus aquaticus* to rapidly do multiple cycles of DNA amplification.

***Polypeptide chain*:** This term refers to a string of amino acids that are linked together by peptide bonds between the amino (N-) terminal and carboxyl (C-) terminal. A polypeptide is another word for a protein.

***Polyubiquitin chain*:** The protein ubiquitin is covalently attached to many different proteins via specific lysine residues. In some cases, a second ubiquitin is then covalently attached to the first ubiquitin and additional ubiquitins are linked to the previous ones until a chain of ubiquitins is attached to the protein, usually marking it for degradation in the proteasome.

***Primary cilium*:** A sensory organelle in some types of non-mitotic cells that is comprised of a microtubule core known as the axoneme. Upon nutrient deprivation, the primary cilium begins to form from the mother centriole and then migrates from the cell center to extend outward from the plasma membrane.

Primer: A relatively short sequence of DNA nucleotides that is used as a starting point for synthesis of a strand of template DNA using the enzyme DNA polymerase. It is needed for DNA amplification by PCR.

Promoter: A specific sequence of DNA that interacts with proteins and regulates the transcription of neighboring DNA sequences to RNA. Promoters are essential parts of plasmids so that the cDNA inserted in the plasmid will be expressed in cells.

Proteasomal degradation: This refers to the proteasomal degradation pathway, where proteins marked by ubiquitin (especially polyubiquitin chains) are targeted to the proteasome where they undergo degradation.

Proteasome: A barrel-like proteinaceous structure that recognizes ubiquitin chains and thus channels polyubiquitinated proteins inside where they undergo degradation into amino acids.

Protein: A series of amino acids linked together in a chain (also known as a polypeptide because the chemical bonds between the amino acids are known as peptide bonds). Proteins are coded for by DNA (genes), which is first transcribed into mRNA and then translated into protein at the ribosome. Proteins carry out myriad essential cellular functions.

Protooncogenes: A normal cellular gene that when mutated can lead to tumorigenesis of the cell.

Pulse-chase: General name for a variety of techniques in which there is an event for a period of time (pulse), but the event is then stopped and the experiment is continued over time (chase). For example, a ligand is allowed to bind to cell surface receptors for 30 minutes (pulse), but then the remaining unbound ligand is washed away and the amount of internalized receptor-ligand complexes in the cell are observed or measured after an additional 30 minutes, 1 hour, and 2 hours, etc. (chase).

R

Receptor: A protein that is expressed on the cell surface membrane and usually has an external region, a membrane spanning hydrophobic ("water-hating") region, and an internal "tail" within the cell. Receptors are known to transmit signals into the cell when they are bound by an external ligand.

Receptor-mediated endocytosis: The process in which receptors on the cell surface bind to their extracellular ligands and are internalized within small membrane-bound vesicles as they curve inwardly from the surface and are pinched off to form separate internal compartments. Internalized vesicles containing receptors in their lipid bilayers fuse with sorting endosomes, and ultimately the internalized receptors are either shuttled to lysosomes for degradation or returned to the plasma membrane in a process known as recycling.

***Recessive*:** Refers to a gene that will be passed on to the offspring only if the individual has two alleles or copies.

***Recycling*:** There are multiple recycling processes in cells, and an important example is the return of receptors to the cell surface once they have been internalized via receptor-mediated endocytosis.

***Reduction*:** The gain of electrons of a molecule, atom, or ion. The molecule is then said to be in a "reduced state."

***Regression to the mean*:** The statistical idea that in random events dependent on multiple variables, data points that are extreme are usually followed by more moderate ones. This depends on the sample size, with the more samples done, the greater the statistical tendency to revert to the mean.

***Ribosomal RNA (rRNA)*:** This is non-coding RNA, or in other words RNA that is not translated into protein, but instead forms an important part of the ribosomal machinery that helps translate mRNA into protein.

***Ribosome*:** This is a complex organelle comprised of numerous proteins and RNA molecules that serves as the machinery for translation of mRNA into protein in the cell. Ribosomes move along "scanning" mRNA molecules and help convert the genetic instructions in the form of mRNA codons to undergo translation into polypeptide chains as they move along the ribosome.

***Rous sarcoma virus (RSV)*:** An RNA virus that holds its genetic material in the form of RNA rather than DNA and its genes need to be reverse transcribed into DNA before undergoing the typical DNA-to-RNA-to-protein process in host cells. RSV was the first virus identified that can transform normal cells into tumor cells.

S

***Sense strand*:** The DNA double helix is comprised of two strands of DNA: the coding strand (known as the sense strand) that comprises the genes that are transcribed and translated into proteins, and the antisense strand which is used as a template to copy the DNA during the process of replication.

Sickle cell anemia: An inherited disease of erythrocytes or red blood cells, in which mutations cause abnormalities in the red blood cell protein hemoglobin, leading to the cells becoming "sickle-shaped." As a result, these cells die prematurely, causing decreased oxygen supply to the tissues, as well as blockage in the blood vessels that increases susceptibility to infections, high blood pressure, heart problems, and other medical issues.

Side chain model of immunity: Proposed by Paul Ehrlich, this model held that normally cells express multiple types of "side chains" on their surface to bind and take up nutrients, and that some pathogens mimic these nutrients. When challenged by a pathogen, the side chains bind to the toxins rather than the nutrients and thus neutralize the ability of the side chains to supply nutrients to the cell. To overcome this, he proposed that the cell produces additional side chains that are secreted and serve as antitoxins (similar to what we now call antibodies) and in this way there is defense against recurrent infections by the same pathogen.

Signal transduction: This is a general term that refers to the transmission of signals within cells, or transmission of signals from receptors at the plasma membrane to the cell interior. The most well-known mechanisms of signal transduction involve binding of a ligand to a receptor, conformational changes in the receptor, and a cascade of post-translational modifications such as phosphorylation, ubiquitination, etc., that go from protein to protein and often culminate in transcriptional activation of select genes.

Silencing RNA (siRNA): A process in which select mRNA translation into a protein is shut down, leading to eventual loss of expression of that protein. This process can occur naturally in cells, but in the last 20 years has been used by researchers (and clinicians) as a technique to selectively block expression of a specific protein.

Somatic theory: This is the accepted theory for antibody generation. One of the difficult problems for immunologists was to determine how antibodies obtain such incredible diversity, and are capable of binding to every possible antigen. Somatic theory maintains that variations in the antibody binding regions result from changes (somatic rearrangement of the antibody genes) that occur during cell replication.

Sorting: A term used by researchers studying the trafficking and movement of proteins from one location to another within the cell. Sorting is the process by which proteins (and lipids) are segregated to specific regions of an organelle's membrane so that they can be included in membrane-bound vesicles for targeting within the cell when these vesicles bud from the organelle.

Src: A protein with oncogenic potential that is encoded by the oncogene *src*. Src was first identified in the Rous sarcoma virus as the protein needed to transform chicken cells into tumor cells but was later found to have a normal cellular counterpart (*c-src*). Src is a kinase that post-translationally phosphorylates (adds a phosphate moiety) to proteins.

Stop codon: A three-nucleotide sequence (codon) that does not specify inclusion of any amino acid to a nascent polypeptide chain, but instead signals termination of translation.

Symbiosis: A process where two distinct organisms benefit from a proximal relationship with one another.

T

Taq polymerase: This is a DNA polymerase from *Thermus aquaticus*, a type of thermophilic bacteria (one that grows at high temperatures) that is found in hot springs at Yellowstone National Park. This DNA polymerase is incredibly useful for amplifying DNA in repeated cycles of heating and cooling because the enzyme is stable at high temperatures.

Telomerase: Refers to a complex that contains both RNA and proteins and is capable of extending the telomeric sequences which are found at the ends of chromosomes and cannot be copied by other DNA polymerases. Telomerase therefore prevents chromosomes from losing their telomeres after multiple cell divisions, and in some cases excess telomerase activity can lead to oncogenesis of cells.

Telomere: Comprised of a sequence repeat of (typically several thousand) TTAGGG nucleotides, this region protects chromosomes from a variety of detrimental events, including rearrangement, fusing of the chromosome ends, DNA damage, and recombination.

Telomeric repeat amplification protocol (TRAP): A PCR-based amplification method used to measure levels of telomerase activity in a cell lysate.

Template: Refers to a segment of DNA that is used in a PCR reaction for its amplification.

TERT: This is a telomerase enzyme that plays a key role in the replication of telomeres at the end of chromosomes.

Thermus aquaticus: A type of thermophilic bacteria (that grows at high temperatures) that is found in hot springs at Yellowstone National

Park and is the source of the important DNA polymerase enzyme used in PCR known as *Taq polymerase.*

***TracrRNA*:** A region of the bacterial genome involved in the CRISPR pathway, TracrRNA is transcribed and then forms a single hybrid molecule with other transcribed RNA molecules from the CRISPR array that specifically recognize the invading pathogen DNA. This hybrid molecule binds to the Cas9 enzyme and then specifically recognizes the DNA that is homologous to the CRISPR array, where the Cas9 can cut and neutralize it.

***Transcription*:** A process in which DNA that is slated to code for a specific protein is copied into RNA so that the latter can be translated into protein on the ribosome according to its codons.

***Transcription activator-like effector nucleases (TALEN)*:** These are restriction enzymes that can be used to cut specific sequences of DNA and have been co-opted for purposes of genetic engineering.

***Transfection*:** A technique for the introduction of DNA into cells usually for the purpose of its expression. Often this is done by using a plasmid to carry the gene of interest.

***Transfer RNA (tRNA)*:** A non-coding RNA molecule of about 70–90 nucleotides that binds to a specific amino acid. A select tRNA exists for each of the 20 amino acids, and it is the tRNA that recognizes the mRNA codon on the ribosome so that it can deliver the correct amino acid to the growing polypeptide chain.

***Transformation*:** Usually refers to the process of a cell becoming oncogenic or tumorigenic, and undergoing a change so that it continuously replicates.

***Transforming principle*:** Before it was known that DNA was the hereditary material, researchers searched for the transforming principle, or the biological material that could change an organism. Oswald Avery was the discoverer of DNA as the transforming principle.

***Transgenic mice*:** Mice whose genomic DNA has been modified via genetic engineering to study the function of the gene and its protein product.

***Translation*:** In the process of protein synthesis in cells, translation refers to the generation of a polypeptide based on the mRNA sequence (its codons) attached to the ribosome.

***Translational science*:** Scientific experiments done in the laboratory designed to lead to interventions that improve the health of individuals and communities, including the development of new diagnostic tools and therapies.

***Tumor suppressors*:** Genes whose expression serves to prevent oncogenesis. p53 is one of the best known and well-characterized tumor suppressors.

***Tumor virology*:** The field studying viruses that can cause tumorigenesis in host organisms after infection.

U

***Ubiquitin*:** A small eight kilodalton protein that is used primarily as a tag for degradation in the proteasomal pathway. A series of enzymes covalently attach it to the substrate protein that is slated for degradation, and then a chain of ubiquitins is attached to the initial ubiquitin. The entire protein is then often degraded in the proteasomal pathway.

V

***Vascular endothelial growth factor (VEGF)*:** A protein that promotes the generation and growth of new blood vessels, or angiogenesis.

***Vector*:** This is another word to describe a plasmid, a circular piece of DNA in which a gene can be inserted in order to transfect (introduce) it into cells.

***Vesicle*:** A small spherical structure surrounded by a bilayer of phospholipids. Vesicles are usually 50–80 nanometers in diameter and are involved in the transport of lipids and proteins within and also outside cells.

Z

***Zinc finger nucleases (ZFN)*:** A class of engineered DNA-binding enzymes that cause double-strand breaks in DNA and can be activated at specific sites. These enzymes are used in certain methods of genetic engineering.

***Zygote*:** The resulting cell once a female cell has been fertilized.

Bibliography

Alvarado-Kristensson, M. 2020. "Choreography of the centrosome." *Heliyon* 6(1):e03238. doi:10.1016/j.heliyon.2020.e03238.

Anderson, R. G., M. S. Brown, and J. L. Goldstein. 1977. "Role of the coated endocytic vesicle in the uptake of receptor-bound low density lipoprotein in human fibroblasts." *Cell* 10(3):351–64. doi: 10.1016/0092-8674(77)90022-8.

Avery, O. T., C. M. Macleod, and M. McCarty. 1944. "Studies on the chemical nature of the substance inducing transformation of pneumococcal types : Induction of transformation by a desoxyribonucleic acid fraction isolated from pneumococcus type Iii." *J Exp Med* 79(2):137–58. doi: 10.1084/jem.79.2.137.

Bainton, D. F. 1981. "The discovery of lysosomes." *J Cell Biol* 91(3 Pt 2):66s–76s. doi: 10.1083/jcb.91.3.66s.

Baker, S. J., E. R. Fearon, J. M. Nigro, S. R. Hamilton, A. C. Preisinger, J. M. Jessup, P. vanTuinen, D. H. Ledbetter, D. F. Barker, Y. Nakamura, R. White, and B. Vogelstein. 1989. "Chromosome 17 deletions and p53 gene mutations in colorectal carcinomas." *Science* 244(4901):217–21. doi: 10.1126/science.2649981.

Baltimore, D. 1970. "RNA-dependent DNA polymerase in virions of RNA tumour viruses." *Nature* 226(5252):1209–11. doi: 10.1038/2261209a0.

Barrangou, R., C. Fremaux, H. Deveau, M. Richards, P. Boyaval, S. Moineau, D. A. Romero, and P. Horvath. 2007. "CRISPR provides acquired resistance against viruses in prokaryotes." *Science* 315(5819):1709–12. doi: 10.1126/science.1138140.

Bartel, B., I. Wunning, and A. Varshavsky. 1990. "The recognition component of the N-end rule pathway." *EMBO J* 9(10):3179–89.

Bassett, E. J., M. S. Keith, G. J. Armelagos, D. L. Martin, and A. R. Villanueva. 1980. "Tetracycline-labeled human bone from ancient Sudanese Nubia (A.D. 350)." *Science* 209(4464):1532–4. doi: 10.1126/science.7001623.

Bateson, W. 1894. *Materials for the Study of Variation: Treated with Especial Regard to Discontinuity in the Origin of Species.* London, UK: Macmillan and Co.

Becker, A. J., C. Ea Mc, and J. E. Till. 1963. "Cytological demonstration of the clonal nature of spleen colonies derived from transplanted mouse marrow cells." *Nature* 197:452–4. doi: 10.1038/197452a0.

Beemon, K., and T. Hunter. 1977. "In vitro translation yields a possible Rous sarcoma virus src gene product." *Proc Natl Acad Sci USA* 74(8):3302–6. doi: 10.1073/pnas.74.8.3302.

Beljanski, M., and S. Ochoa. 1958. "Protein biosynthesis by a cell-free bacterial system." *Proc Natl Acad Sci USA* 44(6):494–501. doi: 10.1073/pnas.44.6.494.

Bello, S. M., C. L. Smith, and J. T. Eppig. 2015. "Allele, phenotype and disease data at Mouse Genome Informatics: Improving access and analysis." *Mamm Genome* 26(7–8):285–94. doi: 10.1007/s00335-015-9582-y.

Bentivoglio, M. 1999. "The discovery of the Golgi apparatus." *J Hist Neurosci* 8(2):202–8. doi: 10.1076/jhin.8.2.202.1833.

Bessman, M. J., I. R. Lehman, E. S. Simms, and A. Kornberg. 1958. "Enzymatic synthesis of deoxyribonucleic acid. II. General properties of the reaction." *J Biol Chem* 233(1):171–7.

Bexiga, M. G., and J. C. Simpson. 2013. "Human diseases associated with form and function of the Golgi complex." *Int J Mol Sci* 14(9):18670–81. doi: 10.3390/ijms140918670.

Billroth, C.A.T. 1871. *General Surgical Pathology Therapeutics: in Fifty Lectures. A Text-book for Students and Physicians.* New York, NY, USA: D Appleton and company.

Bjørneboe, M., H. Gormsen, and F. Lundquist. 1947. "Further experimental studies on the role of the plasma cells as antibody producers." *J Immunol* 55(2):121–9.

Blandino, G., and S. Di Agostino. 2018. "New therapeutic strategies to treat human cancers expressing mutant p53 proteins." *J Exp Clin Cancer Res* 37(1):30. doi: 10.1186/s13046-018-0705-7.

Bliss, M. 1982. *The Discovery of Insulin.* Chicago, IL: The University of Chicago Press.

Bloodgood, R. A. 2009. "From central to rudimentary to primary: The history of an underappreciated organelle whose time has come. The primary cilium." *Methods Cell Biol* 94:3–52. doi: 10.1016/S0091-679X(08)94001-2.

Bloodgood, R. A. 2010. "Sensory reception is an attribute of both primary cilia and motile cilia." *J Cell Sci* 123(4):505–9. doi: 10.1242/jcs.066308.

Boivin, A., and R. Vendrely. 1947. "Sur le role possible des deux acides nucléiques dans la cellule vivante." *Experientia* 3(1):32–4.

Bolotin, A., B. Quinquis, A. Sorokin, and S. D. Ehrlich. 2005. "Clustered regularly interspaced short palindrome repeats (CRISPRs) have spacers of extrachromosomal origin." *Microbiol (Reading)* 151(8):2551–61. doi: 10.1099/mic.0.28048-0.

Boveri, T. 2008. "Concerning the origin of malignant tumours by Theodor Boveri. Translated and annotated by Henry Harris. " Translated and annotated by Henry Harris. *J Cell Sci* 121(Suppl 1):1–84. doi: 10.1242/jcs.025742.

Bowen, R. H. 1923. "The origin of secretory granules." *Proc Natl Acad Sci USA* 9(10):349–52. doi: 10.1073/pnas.9.10.349.

Box, George E. P. 1979. "Robustness in the strategy of scientific model building." In: *Robustness in Statistics*, edited by Robert L. Launer and Graham N. Wilkinson , 201–36. Academic Press.

Boylston, A. 2012. "The origins of inoculation." *J R Soc Med* 105(7):309–13. doi: 10.1258/jrsm.2012.12k044.

Brennan, F. M., D. Chantry, A. Jackson, R. Maini, and M. Feldmann. 1989. "Inhibitory effect of TNF alpha antibodies on synovial cell interleukin-1 production in rheumatoid arthritis." *Lancet* 2(8657):244–7. doi: 10.1016/s0140-6736(89)90430-3.

Brock, T. D. 1997. "The value of basic research: Discovery of Thermus aquaticus and other extreme thermophiles." *Genetics* 146(4):1207–10.

Brouns, S. J., M. M. Jore, M. Lundgren, E. R. Westra, R. J. Slijkhuis, A. P. Snijders, M. J. Dickman, K. S. Makarova, E. V. Koonin, and J. van der Oost. 2008. "Small CRISPR RNAs guide antiviral defense in prokaryotes." *Science* 321(5891):960–4. doi: 10.1126/science.1159689.

Brown, R., C. J. Marshall, S. G. Pennie, and A. Hall. 1984. "Mechanism of activation of an N-ras gene in the human fibrosarcoma cell line HT1080." *EMBO J* 3(6):1321–6.

Brugge, J. S., and R. L. Erikson. 1977. "Identification of a transformation-specific antigen induced by an avian sarcoma virus." *Nature* 269(5626):346–8. doi: 10.1038/269346a0.

Burgos, M. H., and D. W. Fawcett. 1955. "Studies on the fine structure of the mammalian testis. I. Differentiation of the spermatids in the cat (Felis domestica)." *J Biophys Biochem Cytol* 1(4):287–300. doi: 10.1083/jcb.1.4.287.

Burnet, F. M. 1957. "A modification of Jerne's theory of antibody production using the concept of clonal selection." *Aust J Sci* 20:67–9.

Bykov, V. J., N. Issaeva, A. Shilov, M. Hultcrantz, E. Pugacheva, P. Chumakov, J. Bergman, K. G. Wiman, and G. Selivanova. 2002. "Restoration of the tumor suppressor function to mutant p53 by a low-molecular-weight compound." *Nat Med* 8(3):282–8. doi: 10.1038/nm0302-282.

Cancer Genome Atlas Research Network. 2011. "Integrated genomic analyses of ovarian carcinoma." *Nature* 474(7353):609–15. doi: 10.1038/nature10166.

Capecchi, M. R. 1994. "Targeted gene replacement." *Sci Am* 270(3):52–9. doi: 10.1038/scientificamerican0394-52.

Caplan, S. 2012. "Coming out of the scientific closet: Unapologetic about basic research." *The Guardian*, December 11, 2012. https://www.theguardian.com/science/occams-corner/2012/dec/11/scientific-closet-basic-research. Occam's Corner.

Caplan, S., O. Almogi-Hazan, A. Ezernitchi, E. Manaster, A. Gazit, and M. Baniyash. 2001. "The cytoskeleton-associated TCR zeta chain is constitutively phosphorylated in the absence of an active p56(lck) form." *Eur J Immunol* 31(2):580–9. doi:10.1002/1521-4141(200102)31:2<580::aid-immu580>3.0.co;2-h.

Caplan, S., and J. S. Bonifacino. 2003. "Lysosomes." In: *Intracellular Pathogens in Membrane Interactions and Vacuole Biogenesis*. Edited by Jean-Pierre Gorvel, 16–33. Landes Bioscience, Kluwer Academic/Plenum Publishers.

Caplan, S., and M. Baniyash. 1995. "Multisubunit receptors in the immune system and their association with the cytoskeleton: In search of functional significance." *Immunol Res* 14(2):98–118. doi: 10.1007/BF02918171.

Caplan, S., and M. Baniyash. 1996. "Normal T cells express two T cell antigen receptor populations, one of which is linked to the cytoskeleton via zeta chain and displays a unique activation-dependent phosphorylation pattern." *J Biol Chem* 271(34):20705–12. doi: 10.1074/jbc.271.34.20705.

Caplan, S., and M. Baniyash. 2000. "Searching for significance in TCR-cytoskeleton interactions." *Immunol Today* 21(5):223–8. doi: 10.1016/s0167-5699(00)01604-2.

Caplan, S., E. C. Dell'Angelica, W. A. Gahl, and J. S. Bonifacino. 2000. "Trafficking of major histocompatibility complex class II molecules in human B-lymphoblasts deficient in the AP-3 adaptor complex." *Immunol Lett* 72(2):113–7. doi: 10.1016/s0165-2478(00)00176-0.

Caplan, S., R. Gallily, and Y. Barenholz. 1994. "Characterization and purification of a mycoplasma membrane-derived macrophage-activating factor." *Cancer Immunol Immunother* 39(1):27–33. doi: 10.1007/BF01517177.

Caplan, S., L. M. Hartnell, R. C. Aguilar, N. Naslavsky, and J. S. Bonifacino. 2001. "Human Vam6p promotes lysosome clustering and fusion in vivo." *J Cell Biol* 154(1):109–22. doi: 10.1083/jcb.200102142.

Caplan, S., N. Naslavsky, L. M. Hartnell, R. Lodge, R. S. Polishchuk, J. G. Donaldson, and J. S. Bonifacino. 2002. "A tubular EHD1-containing compartment involved in the recycling of major histocompatibility complex class I molecules to the plasma membrane." *EMBO J* 21(11):2557–67. doi: 10.1093/emboj/21.11.2557.

Caplan, S., S. Zeliger, L. Wang, and M. Baniyash. 1995. "Cell-surface-expressed T-cell antigen-receptor zeta chain is associated with the cytoskeleton." *Proc Natl Acad Sci USA* 92(11):4768–72. doi: 10.1073/pnas.92.11.4768.

Carpenter, G., and S. Cohen. 1976. "125I-labeled human epidermal growth factor. Binding, internalization, and degradation in human fibroblasts." *J Cell Biol* 71(1):159–71. doi: 10.1083/jcb.71.1.159.

Chalfie, M. 2009. "GFP: Lighting up life." *Proc Natl Acad Sci USA* 106(25):10073–80. doi: 10.1073/pnas.0904061106.

Chalfie, M., Y. Tu, G. Euskirchen, W. W. Ward, and D. C. Prasher. 1994. "Green fluorescent protein as a marker for gene expression." *Science* 263(5148):802–5. doi: 10.1126/science.8303295.

Chambers, R. 1924. *General Cytology.* Edited by E. V. Cowdry. Chicago, IL: University of Chicago Press.

Chapman, M. J. 1998. "One hundred years of centrioles: The Henneguy-Lenhossek theory, meeting report." *Int Microbiol* 1(3):233–6.

Chau, V., J. W. Tobias, A. Bachmair, D. Marriott, D. J. Ecker, D. K. Gonda, and A. Varshavsky. 1989. "A multiubiquitin chain is confined to specific lysine in a targeted short-lived protein." *Science* 243(4898):1576–83. doi: 10.1126/science.2538923.

Chung, C. 2019. "Current targeted therapies in lymphomas." *Am J Health Syst Pharm* 76(22):1825–34. doi: 10.1093/ajhp/zxz202.

Ciechanover, A. 2005. "Early work on the ubiquitin proteasome system, an interview with Aaron Ciechanover. Interview by CDD." *Cell Death Differ* 12(9):1167–77.

Ciechanover, A., D. Finley, and A. Varshavsky. 1984. "The ubiquitin-mediated proteolytic pathway and mechanisms of energy-dependent intracellular protein degradation." *J Cell Biochem* 24(1):27–53. doi: 10.1002/jcb.240240104.

Ciechanover, A., H. Heller, S. Elias, A. L. Haas, and A. Hershko. 1980. "ATP-dependent conjugation of reticulocyte proteins with the polypeptide required for protein degradation." *Proc Natl Acad Sci USA* 77(3):1365–8. doi: 10.1073/pnas.77.3.1365.

Cobb, M. 2014. "Oswald Avery, DNA, and the transformation of biology." *Curr Biol* 24(2):R55–60. doi: 10.1016/j.cub.2013.11.060.

Cobb, M. 2015. "Who discovered messenger RNA?." *Curr Biol* 25(13):R526–32. doi: 10.1016/j.cub.2015.05.032.

Cong, L., F. A. Ran, D. Cox, S. Lin, R. Barretto, N. Habib, P. D. Hsu, X. Wu, W. Jiang, L. A. Marraffini, and F. Zhang. 2013. "Multiplex genome engineering using CRISPR/Cas systems." *Science* 339(6121):819–23. doi: 10.1126/science.1231143.

Counter, C. M., H. W. Hirte, S. Bacchetti, and C. B. Harley. 1994. "Telomerase activity in human ovarian carcinoma." *Proc Natl Acad Sci USA* 91(8):2900–4. doi: 10.1073/pnas.91.8.2900.

Cowdry, E. V. 1921. "Flagellated thyroid cells in the dogfish (Mustelus canis)." *Anat Rec* 22(5):289–99.

Cox, A. D., and C. J. Der. 2010. "Ras history: The saga continues." *Small GTPases* 1(1):2–27. doi: 10.4161/sgtp.1.1.12178.

Crawford, L. 1983. "The 53,000-dalton cellular protein and its role in transformation." *Int Rev Exp Pathol* 25:1–50.

Cross, M. J., and L. Claesson-Welsh. 2001. "FGF and VEGF function in angiogenesis: Signalling pathways, biological responses and therapeutic inhibition." *Trends Pharmacol Sci* 22(4):201–7. doi: 10.1016/s0165-6147(00)01676-x.

Culotta, C. A. 1970. "Tissue oxidation and theoretical physiology: Bernard, Ludwig, and Pflüger." *Bull Hist Med* 44(2):109–40.

Cyranoski, D. 2019. "The CRISPR-baby scandal: What's next for human gene-editing." *Nature* 566(7745):440–2. doi: 10.1038/d41586-019-00673-1.

Dagan, N., N. Barda, E. Kepten, O. Miron, S. Perchik, M. A. Katz, M. A. Hernan, M. Lipsitch, B. Reis, and R. D. Balicer. 2021. "BNT162b2 mRNA Covid-19 vaccine in a nationwide mass vaccination setting." *N Engl J Med.* doi: 10.1056/NEJMoa2101765.

Dahm, R. 2005. "Friedrich Miescher and the discovery of DNA." *Dev Biol* 278(2):274–88. doi: 10.1016/j.ydbio.2004.11.028.

Dalton, A. J., and M. D. Felix. 1954. "Cytologic and cytochemical characteristics of the Golgi substance of epithelial cells of the epididymis in situ, in homogenates and after isolation." *Am J Anat* 94(2):171–207. doi: 10.1002/aja.1000940202.

Dalton, A. J., and M. D. Felix. 1956. "A comparative study of the Golgi complex." *J Biophys Biochem Cytol* 2(4, Suppl):79–84. doi: 10.1083/jcb.2.4.79.

Dammermann, A., H. Pemble, B. J. Mitchell, I. McLeod, J. R. Yates, 3rd, C. Kintner, A. B. Desai, and K. Oegema. 2009. "The hydrolethalus syndrome protein HYLS-1 links core centriole structure to cilia formation." *Genes Dev* 23(17):2046–59. doi: 10.1101/gad.1810409.

Darwin, C. 1839. *Narrative of the Surveying Voyages of His Majesty's Ships Adventure and Beagle between the Years 1826 and 1836, Describing Their Examination of the Southern Shores of South America, and the Beagle's Circumnavigation of the Globe.* Vol. Journal and remarks. London, UK: Henry Colburn, Great Marlborough Street.

Darwin, C. 1859. *On the Origin of Species. Or the Preservation of Favoured Races in the Struggle for Life.* Albemarle Street, London, UK: John Murray.

de Vries, C., J. A. Escobedo, H. Ueno, K. Houck, N. Ferrara, and L. T. Williams. 1992. "The fms-like tyrosine kinase, a receptor for vascular endothelial growth factor." *Science* 255(5047):989–91. doi: 10.1126/science.1312256.

deHarven, E., and W. Bernhard. 1956. "Etude Au microscope de l'ultrastructure du centriole chez les vertebres." *Mikrosk Anat* 45:378–98.

Deinsberger, J., D. Reisinger, and B. Weber. 2020. "Global trends in clinical trials involving pluripotent stem cells: A systematic multi-database analysis." *NPJ Regen Med* 5:15. doi: 10.1038/s41536-020-00100-4.

DeLeo, A. B., G. Jay, E. Appella, G. C. Dubois, L. W. Law, and L. J. Old. 1979. "Detection of a transformation-related antigen in chemically induced sarcomas and other transformed cells of the mouse." *Proc Natl Acad Sci USA* 76(5):2420–4. doi: 10.1073/pnas.76.5.2420.

Dell'Angelica, E. C., C. Mullins, S. Caplan, and J. S. Bonifacino. 2000. "Lysosome-related organelles." *FASEB J* 14(10):1265–78. doi: 10.1096/fj.14.10.1265.

Deltcheva, E., K. Chylinski, C. M. Sharma, K. Gonzales, Y. Chao, Z. A. Pirzada, M. R. Eckert, J. Vogel, and E. Charpentier. 2011. "CRISPR RNA maturation by trans-encoded small RNA and host factor RNase III." *Nature* 471(7340):602–7. doi: 10.1038/nature09886.

Der, C. J., T. G. Krontiris, and G. M. Cooper. 1982. "Transforming genes of human bladder and lung carcinoma cell lines are homologous to the ras genes of Harvey and Kirsten sarcoma viruses." *Proc Natl Acad Sci USA* 79(11):3637–40. doi: 10.1073/pnas.79.11.3637.

Doyal, L., and T. Muinzer. 2011. "Should the skeleton of "the Irish giant" be buried at sea?" *BMJ* 343:d7597. doi: 10.1136/bmj.d7597.

Doyle, A., M. P. McGarry, N. A. Lee, and J. J. Lee. 2012. "The construction of transgenic and gene knockout/knockin mouse models of human disease." *Transgen Res* 21(2):327–49. doi: 10.1007/s11248-011-9537-3.

Duesberg, P. H., and P. K. Vogt. 1970. "Differences between the ribonucleic acids of transforming and nontransforming avian tumor viruses." *Proc Natl Acad Sci USA* 67(4):1673–80. doi: 10.1073/pnas.67.4.1673.

Durst, M., L. Gissmann, H. Ikenberg, and H. zur Hausen. 1983. "A papillomavirus DNA from a cervical carcinoma and its prevalence in cancer biopsy samples from different geographic regions." *Proc Natl Acad Sci USA* 80(12):3812–5. doi: 10.1073/pnas.80.12.3812.

Duve, C. 1975. "Exploring cells with a centrifuge." *Science* 189(4198):186–94. doi: 10.1126/science.1138375.

Ecker, A. 1844. "Flimmerbewegung im Gehörorgan von Petromyzon marinus." *Arch Anat Physiol Wiss Med* (*Müller's Archiv*) :520–521.

Edelman, G. M. 1973. "Antibody structure and molecular immunology." *Science* 180(4088):830–40. doi: 10.1126/science.180.4088.830.

Elbashir, S. M., J. Harborth, W. Lendeckel, A. Yalcin, K. Weber, and T. Tuschl. 2001. "Duplexes of 21-nucleotide RNAs mediate RNA interference in cultured mammalian cells." *Nature* 411(6836):494–8. doi: 10.1038/35078107.

Ellis, R. W., D. Defeo, T. Y. Shih, M. A. Gonda, H. A. Young, N. Tsuchida, D. R. Lowy, and E. M. Scolnick. 1981. "The p21 src genes of Harvey and Kirsten sarcoma viruses originate from divergent members of a family of normal vertebrate genes." *Nature* 292(5823):506–11. doi: 10.1038/292506a0.

Ernster, L., and G. Schatz. 1981. "Mitochondria: A historical review." *J Cell Biol* 91(3 Pt 2):227s–55s. doi: 10.1083/jcb.91.3.227s.

Esvelt, K. M., A. L. Smidler, F. Catteruccia, and G. M. Church. 2014. "Concerning RNA-guided gene drives for the alteration of wild populations." *eLife* 3. doi: 10.7554/eLife.03401.

Evans, M. J., and M. H. Kaufman. 1981. "Establishment in culture of pluripotential cells from mouse embryos." *Nature* 292(5819):154–6. doi: 10.1038/292154a0.

Fagraeus, A. 1948. "The plasma cellular reaction and its relation to the formation of antibodies in vitro." *J Immunol* 58(1):1–13.

Farquhar, M. G., and G. E. Palade. 1981. "The Golgi apparatus (complex)-(1954–1981)-from artifact to center stage." *J Cell Biol* 91(3 Pt 2):77s–103s. doi: 10.1083/jcb.91.3.77s.

Farquhar, M. G., and S. R. Wellings. 1957. "Electron microscopic evidence suggesting secretory granule formation within the Golgi apparatus." *J Biophys Biochem Cytol* 3(2):319–22. doi: 10.1083/jcb.3.2.319.

Feldmann, M., and R. N. Maini. 2003. "Lasker clinical medical research award. TNF defined as a therapeutic target for rheumatoid arthritis and other autoimmune diseases." *Nat Med* 9(10):1245–50. doi: 10.1038/nm939.

Ferrara, N., and W. J. Henzel. 1989. "Pituitary follicular cells secrete a novel heparin-binding growth factor specific for vascular endothelial cells." *Biochem Biophys Res Commun* 161(2):851–8. doi: 10.1016/0006-291x(89)92678-8.

Finlay, C. A., P. W. Hinds, and A. J. Levine. 1989. "The p53 proto-oncogene can act as a suppressor of transformation." *Cell* 57(7):1083–93. doi: 10.1016/0092-8674(89)90045-7.

Finley, D., A. Ciechanover, and A. Varshavsky. 1984. "Thermolability of ubiquitin-activating enzyme from the mammalian cell cycle mutant ts85." *Cell* 37(1):43–55. doi: 10.1016/0092-8674(84)90299-x.

Fire, A., S. Xu, M. K. Montgomery, S. A. Kostas, S. E. Driver, and C. C. Mello. 1998. "Potent and specific genetic interference by double-stranded RNA in Caenorhabditis elegans." *Nature* 391(6669):806–11. doi: 10.1038/35888.

Fischer, B., A. Dimopoulou, J. Egerer, T. Gardeitchik, A. Kidd, D. Jost, H. Kayserili, Y. Alanay, I. Tantcheva-Poor, E. Mangold, C. Daumer-Haas, S. Phadke, R. I. Peirano, J. Heusel, C. Desphande, N. Gupta, A. Nanda, E. Felix, E. Berry-Kravis, M. Kabra, R. A. Wevers, L. van Maldergem, S. Mundlos, E. Morava, and U. Kornak. 2012. "Further characterization of ATP6V0A2-related autosomal recessive cutis laxa." *Hum Genet* 131(11):1761–73. doi: 10.1007/s00439-012-1197-8.

Fleming, A. 1929. "On the antibacterial action of cultures of a Penicillium, with special reference to their use in the isolation of B. influenzæ." *Br J Exp Pathol.* 10(3):226–36.

Folkman, J. 1971. "Tumor angiogenesis: Therapeutic implications." *N Engl J Med* 285(21):1182–6. doi: 10.1056/NEJM197111182852108.

Folkman, M. J., D. M. Long, Jr., and F. F. Becker. 1962. "Tumor growth in organ culture." *Surg Forum* 13:81–3.

Galton, F. 1886. "Regression towards mediocrity in hereditary stature." *The J Anthropol Institute of Great Britain and Ireland* 15:246–64.

Gan, H. K., A. H. Kaye, and R. B. Luwor. 2009. "The EGFRvIII variant in glioblastoma multiforme." *J Clin Neurosci* 16(6):748–54. doi: 10.1016/j.jocn.2008.12.005.

Garneau, J. E., M. E. Dupuis, M. Villion, D. A. Romero, R. Barrangou, P. Boyaval, C. Fremaux, P. Horvath, A. H. Magadan, and S. Moineau. 2010. "The CRISPR/Cas bacterial immune system cleaves bacteriophage and plasmid DNA." *Nature* 468(7320):67–71. doi: 10.1038/nature09523.

Gasiunas, G., R. Barrangou, P. Horvath, and V. Siksnys. 2012. "Cas9-crRNA ribonucleoprotein complex mediates specific DNA cleavage for adaptive immunity in bacteria." *Proc Natl Acad Sci USA* 109(39):E2579–86. doi: 10.1073/pnas.1208507109.

Gatti, R. A., H. J. Meuwissen, H. D. Allen, R. Hong, and R. A. Good. 1968. "Immunological reconstitution of sex-linked lymphopenic immunological deficiency." *Lancet* 2(7583):1366–9. doi: 10.1016/s0140-6736(68)92673-1.

Gibbs, J. B., I. S. Sigal, M. Poe, and E. M. Scolnick. 1984. "Intrinsic GTPase activity distinguishes normal and oncogenic ras p21 molecules." *Proc Natl Acad Sci USA* 81(18):5704–8. doi: 10.1073/pnas.81.18.5704.

Gillham, N. W. 2001. "Evolution by jumps: Francis Galton and William Bateson and the mechanism of evolutionary change." *Genetics* 159(4):1383–92.

Glotzer, M., A. W. Murray, and M. W. Kirschner. 1991. "Cyclin is degraded by the ubiquitin pathway." *Nature* 349(6305):132–8. doi: 10.1038/349132a0.

Goldfarb, M., K. Shimizu, M. Perucho, and M. Wigler. 1982. "Isolation and preliminary characterization of a human transforming gene from T24 bladder carcinoma cells." *Nature* 296(5856):404–9. doi: 10.1038/296404a0.

Goldmann, E. 1908. "The growth of malignant disease in man and the lower animals, with special reference to the Vascular System." *Proc R Soc Med* 1(Surg Sect):1–13.

Goldstein, J. L., and M. S. Brown. 1989. "Familial hyper-cholesterolemia." In: *The Metabolic Basis of Inherited Disease*, edited by C. R. Scriver, A. L. Beaudet, W.S. Sly and D. valle, 6th ed., 1215–50. New York: McGraw-Hill.

Golgi, C. 1989a. "On the structure of nerve cells. 1898.". *J Microsc* 155(1):3–7. doi: 10.1111/j.1365-2818.1989.tb04294.x.

Golgi, C. 1989b. "On the structure of the nerve cells of the spinal ganglia. 1898.". *J Microsc* 155(1):9–14. doi: 10.1111/j.1365-2818.1989.tb04295.x.

Gonzalez-Suarez, E., C. Geserick, J. M. Flores, and M. A. Blasco. 2005. "Antagonistic effects of telomerase on cancer and aging in K5-mTert transgenic mice." *Oncogene* 24(13):2256–70. doi: 10.1038/sj.onc.1208413.

Gormley, M. 2007. "The first 'molecular disease': A story of Linus Pauling, the intellectual patron." *Endeavour* 31(2):71–7.

Gray, M. W. 2017. "Lynn Margulis and the endosymbiont hypothesis: 50 years later." *Mol Biol Cell* 28(10):1285–7. doi: 10.1091/mbc.E16-07-0509.

Greene, H. S. 1941. "Heterologous transplantation of mammalian tumors : I. The transfer of rabbit tumors to alien species." *J Exp Med* 73(4):461–74. doi: 10.1084/jem.73.4.461.

Greider, C. W., and E. H. Blackburn. 1985. "Identification of a specific telomere terminal transferase activity in Tetrahymena extracts." *Cell* 43(2 Pt 1):405–13. doi: 10.1016/0092-8674(85)90170-9.

Greider, C. W., and E. H. Blackburn. 1989. "A telomeric sequence in the RNA of Tetrahymena telomerase required for telomere repeat synthesis." *Nature* 337(6205):331–7. doi: 10.1038/337331a0.

Griffith, F. 1928. "The significance of pneumococcal types." *J Hyg (Lond)* 27(2):113–59. doi: 10.1017/s0022172400031879.

Gross, C. P., and K. A. Sepkowitz. 1998. "The myth of the medical breakthrough: Smallpox, vaccination, and Jenner reconsidered." *Int J Infect Dis* 3(1):54–60. doi: 10.1016/s1201-9712(98)90096-0.

Guo, S., and K. J. Kemphues. 1995. "par-1, a gene required for establishing polarity in C. elegans embryos, encodes a putative Ser/Thr kinase that is asymmetrically distributed." *Cell* 81(4):611–20. doi: 10.1016/0092-8674(95)90082-9.

Gurdon, J. B. 1962a. "Adult frogs derived from the nuclei of single somatic cells." *Dev Biol* 4:256–73. doi: 10.1016/0012-1606(62)90043-x.

Gurdon, J. B. 1962b. "The developmental capacity of nuclei taken from intestinal epithelium cells of feeding tadpoles." *J Embryol Exp Morphol* 10:622–40.

Gurdon, J. B. 1962c. "The transplantation of nuclei between two species of Xenopus." *Dev Biol* 5:68–83. doi: 10.1016/0012-1606(62)90004-0.

Guyer, R. L., and D. E. Koshland, Jr. 1989. "The molecule of the year." *Science* 246(4937):1543–6. doi: 10.1126/science.2688087.

Hall, B., A. Limaye, and A. B. Kulkarni. 2009. "Overview: Generation of gene knockout mice." *Curr Protoc Cell Biol* Chapter 19:Unit 19 12 19 12:1–17. doi: 10.1002/0471143030.cb1912s44.

Hart, P. D. 1968. "Mycobacterium tuberculosis in macrophages: Effect of certain surfactants and other membrane-active compounds." *Science* 162(3854):686–9. doi: 10.1126/science.162.3854.686.

Harvey, J. J. 1964. "An unidentified virus which causes the rapid production of tumours in mice." *Nature* 204:1104–5. doi: 10.1038/2041104b0.

Hayflick, L., and P. S. Moorhead. 1961. "The serial cultivation of human diploid cell strains." *Exp Cell Res* 25:585–621. doi: 10.1016/0014-4827(61)90192-6.

Helmenstine, A. M. 2020. "Francesco Redi: Founder of experimental biology." *Thoughtco*. https://www.thoughtco.com/biography-of-francesco-redi-4126774.

Henig, R. M. 2000. *The Monk in the Garden*. Boston, MA: Mariner Books.

Hershko, A., and A. Ciechanover. 1982. "Mechanisms of intracellular protein breakdown." *Annu Rev Biochem* 51:335–64. doi: 10.1146/annurev.bi.51.070182.002003.

Hershko, A., D. Ganoth, J. Pehrson, R. E. Palazzo, and L. H. Cohen. 1991. "Methylated ubiquitin inhibits cyclin degradation in clam embryo extracts." *J Biol Chem* 266(25):16376–9.

Hill, M., and J. Hillova. 1972. "Virus recovery in chicken cells tested with Rous sarcoma cell DNA." *Nat New Biol* 237(71):35–9. doi: 10.1038/newbio237035a0.

Hirakawa, M. P., R. Krishnakumar, J. A. Timlin, J. P. Carney, and K. S. Butler. 2020. "Gene editing and CRISPR in the clinic: Current and future perspectives." *Biosci Rep* 40(4):1–37. doi: 10.1042/BSR20200127.

His, W. 1897. "Einleitung". In: *Die Histochemischen und Physiologischen Arbeiten von Friedrich Miescher*. Edited by His, W., et al., Vol. 1, 1–4, Leipzig: F. C. W. Vogel.

Hoffman, R. M. 2015. "Application of GFP imaging in cancer." *Lab Investig* 95(4):432–52. doi: 10.1038/labinvest.2014.154.

Hogeboom, G. H., A. Claude, and R. D. Hotch-Kiss. 1946. "The distribution of cytochrome oxidase and succinoxidase in the cytoplasm of the mammalian liver cell." *J Biol Chem* 165(2):615–29.

Homewood, C. A., D. C. Warhurst, W. Peters, and V. C. Baggaley. 1972. "Lysosomes, pH and the anti-malarial action of chloroquine." *Nature* 235(5332): 50–2. doi: 10.1038/235050a0.

Hooke, R. 1665. *Micrographia, or Some Physiological Descriptions of Minute Bodies Made by Magnifying Glasses, with Observations and Inquiries Thereupon.* London, United Kingdom: Martyn and Allestry, Printers to the Royal Society.

Hozumi, N., and S. Tonegawa. 1976. "Evidence for somatic rearrangement of immunoglobulin genes coding for variable and constant regions." *Proc Natl Acad Sci USA* 73(10):3628–32. doi: 10.1073/pnas.73.10.3628.

Hurley, J. B., M. I. Simon, D. B. Teplow, J. D. Robishaw, and A. G. Gilman. 1984. "Homologies between signal transducing G proteins and ras gene products." *Science* 226(4676):860–2. doi: 10.1126/science.6436980.

Ingram, V. M. 1956. "A specific chemical difference between the globins of normal human and sickle-cell anaemia haemoglobin." *Nature* 178(4537):792–4. doi: 10.1038/178792a0.

Ingram, V. M. 1957. "Gene mutations in human haemoglobin: The chemical difference between normal and sickle cell haemoglobin." *Nature* 180(4581):326–8. doi: 10.1038/180326a0.

Inoue, K. 2019. "Pelizaeus-Merzbacher disease: Molecular and cellular pathologies and associated phenotypes." *Adv Exp Med Biol* 1190:201–16. doi: 10.1007/978-981-32-9636-7_13.

Inouye, S., and F. I. Tsuji. 1994. "Evidence for redox forms of the Aequorea green fluorescent protein." *FEBS Lett* 351(2):211–4. doi: 10.1016/0014-5793(94)00859-0.

Itano, H. A., and L. Pauling. 1949. "A rapid diagnostic test for sickle cell anemia." *Blood* 4(1):66–8.

James, C. R. 2014. *Science Unshackled.* Johns Hopkins University Press.

Jamieson, J. D., and G. E. Palade. 1967. "Intracellular transport of secretory proteins in the pancreatic exocrine cell. II. Transport to condensing vacuoles and zymogen granules." *J Cell Biol* 34(2):597–615. doi: 10.1083/jcb.34.2.597.

Janes, M. R., J. Zhang, L. S. Li, R. Hansen, U. Peters, X. Guo, Y. Chen, A. Babbar, S. J. Firdaus, L. Darjania, J. Feng, J. H. Chen, S. Li, S. Li, Y. O. Long, C. Thach, Y. Liu, A. Zarieh, T. Ely, J. M. Kucharski, L. V. Kessler, T. Wu, K. Yu, Y. Wang, Y. Yao, X. Deng, P. P. Zarrinkar, D. Brehmer, D. Dhanak, M. V. Lorenzi, D. Hu-Lowe, M. P. Patricelli, P. Ren, and Y. Liu. 2018. "Targeting KRAS mutant cancers with a covalent G12C-specific inhibitor." *Cell* 172(3):578–589 e17. doi: 10.1016/j.cell.2018.01.006.

Jenner, E. 1801. "On the origin of the vaccine inoculation." *Med Phys J* 5(28):505–8.

Jenner, E. 1798. *An Inquiry Into the Causes and Effects of the Variolæ Vaccinæ, or Cow-Pox.*

Jerne, N. K. 1955. "The natural-selection theory of antibody formation." *Proc Natl Acad Sci USA* 41(11):849–57. doi: 10.1073/pnas.41.11.849.

Jinek, M., K. Chylinski, I. Fonfara, M. Hauer, J. A. Doudna, and E. Charpentier. 2012. "A programmable dual-RNA-guided DNA endonuclease in adaptive bacterial immunity." *Science* 337(6096):816–21. doi: 10.1126/science.1225829.

Jones, P. T., P. H. Dear, J. Foote, M. S. Neuberger, and G. Winter. 1986. "Replacing the complementarity-determining regions in a human antibody with those from a mouse." *Nature* 321(6069):522–5. doi: 10.1038/321522a0.

Kamine, J., and J. M. Buchanan. 1977. "Cell-free synthesis of two proteins unique to RNA of transforming virions of Rous sarcoma virus." *Proc Natl Acad Sci USA* 74(5):2011–5. doi: 10.1073/pnas.74.5.2011.

Katzenstein, A. L., and J. Maksem. 1979. "Candidal infection of gastric ulcers. Histology, incidence, and clinical significance." *Am J Clin Pathol* 71(2):137–41. doi: 10.1093/ajcp/71.2.137.

Keck, P. J., S. D. Hauser, G. Krivi, K. Sanzo, T. Warren, J. Feder, and D. T. Connolly. 1989. "Vascular permeability factor, an endothelial cell mitogen related to PDGF." *Science* 246(4935):1309–12. doi: 10.1126/science.2479987.

Keilin, D. 1925. "On cytochrome, a respiratory pigment common to animals, yeast and higher plants." *Proc Roy Soc B* 98:312–39.

Keilin, D., and E. F. Hartree. 1945. "Purification and properties of cytochrome c." *Biochem J* 39(4):289–92. doi: 10.1042/bj0390289.

Keilin, D., and E. C. Slater. 1953. "Cytochrome." *Br Med Bull* 9(2):89–97. doi: 10.1093/oxfordjournals.bmb.a074346.

Kennedy, E. P., and A. L. Lehninger. 1949. "Oxidation of fatty acids and tricarboxylic acid cycle intermediates by isolated rat liver mitochondria." *J Biol Chem* 179(2):957–72.

Kim, N. W., M. A. Piatyszek, K. R. Prowse, C. B. Harley, M. D. West, P. L. Ho, G. M. Coviello, W. E. Wright, S. L. Weinrich, and J. W. Shay. 1994. "Specific association of human telomerase activity with immortal cells and cancer." *Science* 266(5193):2011–5. doi: 10.1126/science.7605428.

King, M. L., Jr. 1968. Remaining awake through a great revolution. Speech delivered at the National Cathedral, Washington, DC., on 31 March 1968. Congressional Record, 9 April 1968.

Kirsten, W. H., and L. A. Mayer. 1967. "Morphologic responses to a murine erythroblastosis virus." *J Natl Cancer Inst* 39(2):311–35.

Kleppe, K., E. Ohtsuka, R. Kleppe, I. Molineux, and H. G. Khorana. 1971. "Studies on polynucleotides. XCVI. Repair replications of short synthetic DNA's as catalyzed by DNA polymerases." *J Mol Biol* 56(2):341–61. doi: 10.1016/0022-2836(71)90469-4.

Kohler, G., and C. Milstein. 1975. "Continuous cultures of fused cells secreting antibody of predefined specificity." *Nature* 256(5517):495–7. doi: 10.1038/256495a0.

Kohler, R. E. 1973. "The background to Otto Warburg's conception of the Atmungsferment." *J Hist Biol* 6:171–92. doi: 10.1007/BF00127607.

Kornberg, A. 1959a. "Basic research, the lifeline of medicine." *Nobelprize.org*. https://www.nobelprize.org/prizes/medicine/1959/kornberg/article/.

Kornberg, A. 1959b. Nobel Prize in Physiology or Medicine award speech. *NobelPrize.org*. https://www.nobelprize.org/prizes/medicine/1959/kornberg/article/.

Kornfeld, S. 2018. "A lifetime of adventures in glycobiology." *Annu Rev Biochem* 87:1–21. doi: 10.1146/annurev-biochem-062917-011911.

Kowalevsky, A. 1877. "Weitere Studien über die Entwicklungsgeschichte des Amphioxus lanceolatus, nebsteinem Beitrage zur Homologie des Nervensystems der Würmer und Wirbelthiere." *Arch mikrosk Anat* 13(1): 181–204.

Krebs, H. A., and W. A. Johnson. 1937. "Metabolism of ketonic acids in animal tissues." *Biochem J* 31(4):645–60. doi: 10.1042/bj0310645.

Kress, M., E. May, R. Cassingena, and P. May. 1979. "Simian virus 40-transformed cells express new species of proteins precipitable by anti-simian virus 40 tumor serum." *J Virol* 31(2):472–83. doi: 10.1128/JVI.31.2.472-483.1979.

Kuczynski, E. A., and A. R. Reynolds. 2020. "Vessel co-option and resistance to anti-angiogenic therapy." *Angiogenesis* 23(1):55–74. doi: 10.1007/s10456-019-09698-6.

Kuczynski, E. A., P. B. Vermeulen, F. Pezzella, R. S. Kerbel, and A. R. Reynolds. 2019. "Vessel co-option in cancer." *Nat Rev Clin Oncol* 16(8):469–93. doi: 10.1038/s41571-019-0181-9.

Lander, E. S. 2016. "The heroes of CRISPR." *Cell* 164(1–2):18–28. doi: 10.1016/j.cell.2015.12.041.

Lane, D. P., and L. V. Crawford. 1979. "T antigen is bound to a host protein in SV40-transformed cells." *Nature* 278(5701):261–3. doi: 10.1038/278261a0.

Langer, R., and J. Folkman. 1976. "Polymers for the sustained release of proteins and other macromolecules." *Nature* 263(5580):797–800. doi: 10.1038/263797a0.

Langerhans, P. 1876. "Zur anatomie des amphioxus." *Mikrok anat* 12:290–348.

Leder, P., B. F. Clark, W. S. Sly, S. Pestka, and M. W. Nirenberg. 1963. "Cell-free peptide synthesis dependent Upon synthetic oligodeoxynucleotides." *Proc Natl Acad Sci USA* 50:1135–43. doi: 10.1073/pnas.50.6.1135.

Lee, R. C., R. L. Feinbaum, and V. Ambros. 1993. "The C. elegans heterochronic gene lin-4 encodes small RNAs with antisense complementarity to lin-14." *Cell* 75(5):843–54. doi: 10.1016/0092-8674(93)90529-y.

Leeuwenhoek, A. 1677. "Concerning little animals observed in rain-, well-, sea-and snow-water; as also in water wherein pepper had lain infused." *Philos. Translator* 12:821–31.

Lehman, I. R., M. J. Bessman, E. S. Simms, and A. Kornberg. 1958. "Enzymatic synthesis of deoxyribonucleic acid. I. Preparation of substrates and partial purification of an enzyme from Escherichia coli." *J Biol Chem* 233(1):163–70.

Lehman, I. R., S. B. Zimmerman, J. Adler, M. J. Bessman, E. S. Simms, and A. Kornberg. 1958. "Enzymatic synthesis of deoxyribonucleic acid. V. Chemical composition of enzymatically synthesized deoxyribonucleic acid." *Proc Natl Acad Sci USA* 44(12):1191–6. doi: 10.1073/pnas.44.12.1191.

Lemmon, M. A., and J. Schlessinger. 2010. "Cell signaling by receptor tyrosine kinases." *Cell* 141(7):1117–34. doi: 10.1016/j.cell.2010.06.011.

Lenzi, P., G. Bocci, and G. Natale. 2016. "John Hunter and the origin of the term "angiogenesis"." *Angiogenesis* 19(2):255–6. doi: 10.1007/s10456-016-9496-7.

Levine, R., and C. Evers. 2008. *The Slow Death of Spontaneous Generation (1668–1859)*. The National Health Museum. https://webprojects.oit.ncsu.edu/project/bio183de/Black/cellintro/cellintro_reading/Spontaneous_Generation.html.

Lewis, W. H. 1936. "Pinocytosis : Drinking by cells / Warren H. Lewis." *Johns Hopkins Med Arch.*

Li, P., D. Nijhawan, I. Budihardjo, S. M. Srinivasula, M. Ahmad, E. S. Alnemri, and X. Wang. 1997. "Cytochrome c and dATP-dependent formation of Apaf-1/caspase-9 complex initiates an apoptotic protease cascade." *Cell* 91(4):479–89. doi: 10.1016/s0092-8674(00)80434-1.

Linzer, D. I., and A. J. Levine. 1979. "Characterization of a 54K dalton cellular SV40 tumor antigen present in SV40-transformed cells and uninfected embryonal carcinoma cells." *Cell* 17(1):43–52. doi: 10.1016/0092-8674(79)90293-9.

Linzer, D. I., W. Maltzman, and A. J. Levine. 1979. "The SV40 A gene product is required for the production of a 54,000 MW cellular tumor antigen." *Virology* 98(2):308–18. doi: 10.1016/0042-6822(79)90554-3.

Littlefield, J. W., E. B. Keller, J. Gross, and P. C. Zamecnik. 1955. "Studies on cytoplasmic ribonucleoprotein particles from the liver of the rat." *J Biol Chem* 217(1):111–23.

Liu, X., C. N. Kim, J. Yang, R. Jemmerson, and X. Wang. 1996. "Induction of apoptotic program in cell-free extracts: Requirement for dATP and cytochrome c." *Cell* 86(1):147–57. doi: 10.1016/s0092-8674(00)80085-9.

Lohmann, K. 1929. "Uber die Pyrophosphatfraktion im Muskel." *Naturwissenschaften* 17(31):624–5.

Maddox, B. 2003. "The double helix and the 'wronged heroine'." *Nature* 421(6921):407–8. doi: 10.1038/nature01399.

Maderspacher, F. 2008. "Theodor Boveri and the natural experiment." *Curr Biol* 18(7):R279–86. doi: 10.1016/j.cub.2008.02.061.

Makarova, K. S., N. V. Grishin, S. A. Shabalina, Y. I. Wolf, and E. V. Koonin. 2006. "A putative RNA-interference-based immune system in prokaryotes: Computational analysis of the predicted enzymatic machinery, functional analogies with eukaryotic RNAi, and hypothetical mechanisms of action." *Biol Direct* 1:7. doi: 10.1186/1745-6150-1-7.

Mali, P., L. Yang, K. M. Esvelt, J. Aach, M. Guell, J. E. DiCarlo, J. E. Norville, and G. M. Church. 2013. "RNA-guided human genome engineering via Cas9." *Science* 339(6121):823–6. doi: 10.1126/science.1232033.

Malumbres, M., and M. Barbacid. 2003. "RAS oncogenes: The first 30 years." *Nat Rev Cancer* 3(6):459–65. doi: 10.1038/nrc1097.

Manning, G., D. B. Whyte, R. Martinez, T. Hunter, and S. Sudarsanam. 2002. "The protein kinase complement of the human genome." *Science* 298(5600):1912–34. doi: 10.1126/science.1075762.

Marcon, A., Z. Master, V. Ravitsky, and T. Caulfield. 2019. "CRISPR in the North American popular press." *Genet Med* 21(10):2184–9. doi: 10.1038/s41436-019-0482-5.

Marraffini, L. A., and E. J. Sontheimer. 2008. "CRISPR interference limits horizontal gene transfer in staphylococci by targeting DNA." *Science* 322(5909):1843–5. doi: 10.1126/science.1165771.

Martin, G. R. 1981. "Isolation of a pluripotent cell line from early mouse embryos cultured in medium conditioned by teratocarcinoma stem cells." *Proc Natl Acad Sci USA* 78(12):7634–8. doi: 10.1073/pnas.78.12.7634.

Martin, G. S. 1970. "Rous sarcoma virus: A function required for the maintenance of the transformed state." *Nature* 227(5262):1021–3. doi: 10.1038/2271021a0.

Martin, G. S. 2004. "The road to Src." *Oncogene* 23(48):7910–7. doi: 10.1038/sj.onc.1208077.

Martin, R. G., J. H. Matthaei, O. W. Jones, and M. W. Nirenberg. 1962. "Ribonucleotide composition of the genetic code." *Biochem Biophys Res Commun* 6:410–4. doi: 10.1016/0006-291x(62)90365-0.

Matthaei, H., and M. W. Nirenberg. 1961a. "The dependence of cell-free protein synthesis in E. coli upon RNA prepared from ribosomes." *Biochem Biophys Res Commun* 4(6):404–8. doi: 10.1016/0006-291x(61)90298-4.

Matthaei, J. H., O. W. Jones, R. G. Martin, and M. W. Nirenberg. 1962. "Characteristics and composition of RNA coding units." *Proc Natl Acad Sci USA* 48:666–77. doi: 10.1073/pnas.48.4.666.

Matthaei, J. H., and M. W. Nirenberg. 1961b. "Characteristics and stabilization of DNAase-sensitive protein synthesis in E. coli extracts." *Proc Natl Acad Sci USA* 47:1580–8. doi: 10.1073/pnas.47.10.1580.

McClintock, B. 1939. "The behavior in successive nuclear divisions of a chromosome broken at meiosis." *Proc Natl Acad Sci USA* 25(8):405–16. doi: 10.1073/pnas.25.8.405.

Meadows, K. L., and H. I. Hurwitz. 2012. "Anti-VEGF therapies in the clinic." *Cold Spring Harb Perspect Med* 2(10):1–27. doi: 10.1101/cshperspect.a006577.

Meehan, T. F., N. Conte, D. B. West, J. O. Jacobsen, J. Mason, J. Warren, C. K. Chen, I. Tudose, M. Relac, P. Matthews, N. Karp, L. Santos, T. Fiegel, N. Ring, H. Westerberg, S. Greenaway, D. Sneddon, H. Morgan, G. F. Codner, M. E. Stewart, J. Brown, N. Horner, Consortium International Mouse Phenotyping, M. Haendel, N. Washington, C. J. Mungall, C. L. Reynolds, J. Gallegos, V. Gailus-Durner, T. Sorg, G. Pavlovic, L. R. Bower, M. Moore, I. Morse, X. Gao, G. P. Tocchini-Valentini, Y. Obata, S. Y. Cho, J. K. Seong, J. Seavitt, A. L. Beaudet, M. E. Dickinson, Y. Herault, W. Wurst, M. H. de Angelis, K. C. K. Lloyd, A. M. Flenniken, L. M. J. Nutter, S. Newbigging, C. McKerlie, M. J. Justice, S. A. Murray, K. L. Svenson, R. E. Braun, J. K. White, A. Bradley, P. Flicek, S. Wells, W. C. Skarnes, D. J. Adams, H. Parkinson, A. M. Mallon, S. D. M. Brown, and D. Smedley. 2017. "Disease model discovery from 3,328 gene knockouts by The International Mouse Phenotyping Consortium." *Nat Genet* 49 (8):1231–1238. doi: 10.1038/ng.3901.

Miescher, F. 1870. "Nachtragliche Bemerkungen". In: *Die Histochemischen und Physiologischen Arbeiten von Friedrich Miescher—A. Arbeiten von F. Miescher.* Edited by His, W., et al., Vol. 2, 32–4, Leipzig: F. C. W. Vogel.

Miescher, Friederich. 1871. "Ueber die chemische Zusammensetzung der Eiterzellen." *Med.-Chem. Unters* 4:441–60.

Miller, G., P. Wertheim, G. Wilson, J. Robinson, J. L. Geelen, J. van der Noordaa, and A. J. van der Eb. 1979. "Transfection of human lymphoblastoid cells with herpes simplex viral DNA." *Proc Natl Acad Sci USA* 76(2):949–53. doi: 10.1073/pnas.76.2.949.

Miller, V. J., and D. Ungar. 2012. "Re'COG'nition at the Golgi." *Traffic* 13(7):891–7. doi: 10.1111/j.1600-0854.2012.01338.x.

Minev, B., J. Hipp, H. Firat, J. D. Schmidt, P. Langlade-Demoyen, and M. Zanetti. 2000. "Cytotoxic T cell immunity against telomerase reverse transcriptase in humans." *Proc Natl Acad Sci USA* 97(9):4796–801. doi: 10.1073/pnas.070560797.

Mitchell, P. 1961. "Coupling of phosphorylation to electron and hydrogen transfer by a chemi-osmotic type of mechanism." *Nature* 191:144–8. doi: 10.1038/191144a0.

Mojica, F. J., G. Juez, and F. Rodriguez-Valera. 1993. "Transcription at different salinities of Haloferax mediterranei sequences adjacent to partially modified PstI sites." *Mol Microbiol* 9(3):613–21. doi: 10.1111/j.1365-2958.1993.tb01721.x.

Morin, G. B. 1989. "The human telomere terminal transferase enzyme is a ribonucleoprotein that synthesizes TTAGGG repeats." *Cell* 59(3):521–9. doi: 10.1016/0092-8674(89)90035-4.

Morrison, S. L., M. J. Johnson, L. A. Herzenberg, and V. T. Oi. 1984. "Chimeric human antibody molecules: Mouse antigen-binding domains with human constant region domains." *Proc Natl Acad Sci USA* 81(21):6851–5. doi: 10.1073/pnas.81.21.6851.

Muller, H. J. 1938. "Bar duplication." *Collecting Net (Woods Hole)* 13:181–98.

Muthukkaruppan, V., and R. Auerbach. 1979. "Angiogenesis in the mouse cornea." *Science* 205(4413):1416–8. doi: 10.1126/science.472760.

Napoli, C., C. Lemieux, and R. Jorgensen. 1990. "Introduction of a Chimeric Chalcone synthase Gene into Petunia Results in Reversible Co-Suppression of Homologous Genes in trans." *Plant Cell* 2(4):279–89. doi: 10.1105/tpc.2.4.279.

Nassonov, D. N. 1923. "Stages of Golgi bodies in Protozoa." *Arch MIK Anat* 97:136–86.

Neufeld, E. F. 2011. "From serendipity to therapy." *Annu Rev Biochem* 80:1–15. doi: 10.1146/annurev.biochem.031209.093756.

Neutra, M., and C. P. Leblond. 1966. "Synthesis of the carbohydrate of mucus in the Golgi complex as shown by electron microscope radioautography of goblet cells from rats injected with glucose-H3." *J Cell Biol* 30(1):119–36. doi: 10.1083/jcb.30.1.119.

Nirenberg, M., and P. Leder. 1964. "Rna codewords and protein synthesis. The effect of trinucleotides Upon the binding of srna to ribosomes." *Science* 145(3639):1399–407. doi: 10.1126/science.145.3639.1399.

Nirenberg, M. W., and J. H. Matthaei. 1961. "The dependence of cell-free protein synthesis in E. coli upon naturally occurring or synthetic polyribonucleotides." *Proc Natl Acad Sci USA* 47:1588–602. doi: 10.1073/pnas.47.10.1588.

Nirenberg, M. W., J. H. Matthaei, O. W. Jones, R. G. Martin, and S. H. Barondes. 1963. "Approximation of genetic code via cell-free protein synthesis directed by template RNA." *Fed Proc* 22:55–61.

Nossal, G. J., and J. Lederberg. 1958. "Antibody production by single cells." *Nature* 181(4620):1419–20. doi: 10.1038/1811419a0.

O'Bryan, J. P. 2019. "Pharmacological targeting of RAS: Recent success with direct inhibitors." *Pharmacol Res* 139:503–11. doi: 10.1016/j.phrs.2018.10.021.

Obrist, F., G. Manic, G. Kroemer, I. Vitale, and L. Galluzzi. 2015. "Trial Watch: Proteasomal inhibitors for anticancer therapy." *Mol Cell Oncol* 2(2):e974463. doi: 10.4161/23723556.2014.974463.

Olovnikov, A. M. 1973. "A theory of marginotomy. The incomplete copying of template margin in enzymic synthesis of polynucleotides and biological significance of the phenomenon." *J Theor Biol* 41(1):181–90. doi: 10.1016/0022-5193(73)90198-7.

Palade, G. E. 1953a. "An electron microscope study of the mitochondrial structure." *J Histochem Cytochem* 1(4):188–211. doi: 10.1177/1.4.188.

Palade, G. E. 1953b. "Fine structure of blood capillaries." *J Aoppl Phys* 24:1424.

Palade, G. E. 1955. "A small particulate component of the cytoplasm." *J Biophys Biochem Cytol* 1(1):59–68. doi: 10.1083/jcb.1.1.59.

Parada, L. F., C. J. Tabin, C. Shih, and R. A. Weinberg. 1982. "Human EJ bladder carcinoma oncogene is homologue of Harvey sarcoma virus ras gene." *Nature* 297(5866):474–8. doi: 10.1038/297474a0.

Parada, L. F., and R. A. Weinberg. 1983. "Presence of a Kirsten murine sarcoma virus ras oncogene in cells transformed by 3-methylcholanthrene." *Mol Cell Biol* 3(12):2298–301. doi: 10.1128/mcb.3.12.2298.

Parsons, S. J., and J. T. Parsons. 2004. "Src family kinases, key regulators of signal transduction." *Oncogene* 23(48):7906–9. doi: 10.1038/sj.onc.1208160.

Pauling, L., and H. A. Itano. 1949. "Sickle cell anemia a molecular disease." *Science* 110(2865):543–8. doi: 10.1126/science.110.2865.543.

Pazour, G. J., and G. B. Witman. 2003. "The vertebrate primary cilium is a sensory organelle." *Curr Opin Cell Biol* 15(1):105–10. doi: 10.1016/s0955-0674(02)00012-1.

Perucho, M., M. Goldfarb, K. Shimizu, C. Lama, J. Fogh, and M. Wigler. 1981. "Human-tumor-derived cell lines contain common and different transforming genes." *Cell* 27(3 Pt 2):467–76. doi: 10.1016/0092-8674(81)90388-3.

News, Alex Philippidis senior news Editor Genetic Engineering & Biotechnology. *Gen Hall Shame*. https://www.genengnews.com/a-lists/gen-hall-of-shame/.

Porter, R. R. 1967. "The structure of antibodies. The basic pattern of the principal class of molecules that neutralize antigens (foreign substances in the body) is four cross-linked chains. This pattern is modified so that antibodies can fit different antigens. This pattern is modified so that antibodies can fit different antigens." *Sci Am* 217(4):81–7 passim.

Prasher, D. C., V. K. Eckenrode, W. W. Ward, F. G. Prendergast, and M. J. Cormier. 1992. "Primary structure of the Aequorea victoria green-fluorescent protein." *Gene* 111(2):229–33. doi: 10.1016/0378-1119(92)90691-h.

Pulciani, S., E. Santos, A. V. Lauver, L. K. Long, K. C. Robbins, and M. Barbacid. 1982. "Oncogenes in human tumor cell lines: Molecular cloning of a transforming gene from human bladder carcinoma cells." *Proc Natl Acad Sci USA* 79(9):2845–9. doi: 10.1073/pnas.79.9.2845.

Rabinow, P. 1996. *Making PCR: A Story of Biotechnology*. Chicago, IL: The University of Chicago Press.

Relitti, N., A. P. Saraswati, S. Federico, T. Khan, M. Brindisi, D. Zisterer, S. Brogi, S. Gemma, S. Butini, and G. Campiani. 2020. "Telomerase-based cancer therapeutics: A review on their clinical trials." *Curr Top Med Chem* 20(6):433–57. doi: 10.2174/1568026620666200102104930.

Ribatti, D. 2014. "History of research on angiogenesis." *Chem Immunol Allergy* 99:1–14. doi: 10.1159/000353311.

Ribatti, D. 2016. "The discovery of immunoglobulin E." *Immunol Lett* 171:1–4. doi: 10.1016/j.imlet.2016.01.001.

Ribatti, D., A. Vacca, and M. Presta. 2000. "The discovery of angiogenic factors: A historical review." *Gen Pharmacol* 35(5):227–31. doi: 10.1016/s0306-3623(01)00112-4.

Rivera-Torres, J., and E. San Jose. 2019. "Src Tyrosine Kinase Inhibitors: New Perspectives on Their Immune, Antiviral, and Senotherapeutic Potential." *Front Pharmacol* 10:1011. doi: 10.3389/fphar.2019.01011.

Roberts, R. B. 1958. Microsomal particles and protein synthesis; Papers presented at the First Symposium of the Biophysical Society, at the Massachusetts Institute of Technology, Cambridge. Biophysical Society.

Rosenfeld, L. 2002. "Insulin: Discovery and controversy." *Clin Chem* 48(12):2270–88.

Rosenthal, R. A., J. F. Megyesi, W. J. Henzel, N. Ferrara, and J. Folkman. 1990. "Conditioned medium from mouse sarcoma 180 cells contains vascular endothelial growth factor." *Growth Factors* 4(1):53–9. doi: 10.3109/08977199009011010.

Roth, T. F., and K. R. Porter. 1964. "Yolk protein uptake in the oocyte of the mosquito Aedes aegypti. L." *J Cell Biol* 20:313–32. doi: 10.1083/jcb.20.2.313.

Rous, P. 1911. "A sarcoma of the fowl transmissible by an agent separable from the tumor cells." *J Exp Med* 13(4):397–411. doi: 10.1084/jem.13.4.397.

Rous, P. 1959. "Surmise and fact on the nature of cancer." *Nature* 183(4672):1357–61. doi: 10.1038/1831357a0.

Rous, P., and J. B. Murphy. 1914. "On the causation by filterable agents of three distinct chicken tumors." *J Exp Med* 19(1):52–68. doi: 10.1084/jem.19.1.52.

Rowe, W. P., J. W. Hartley, I. Brodsky, and R. J. Huebner. 1958. "Complement fixation with a mouse tumor virus (S.E. polyoma)." *Science* 128(3335):1339–40. doi: 10.1126/science.128.3335.1339.

Rubin, H. 1955. "Quantitative relations between causative virus and cell in the Rous no. 1 chicken sarcoma." *Virology* 1(5):445–73. doi: 10.1016/0042-6822(55)90037-4.

Ruddell, W. S., A. T. Axon, J. M. Findlay, B. A. Bartholomew, and M. J. Hill. 1980. "Effect of cimetidine on the gastric bacterial flora." *Lancet* 1(8170):672–4.

Sagan, L. 1967. "On the origin of mitosing cells." *J Theor Biol* 14(3):255–74. doi: 10.1016/0022-5193(67)90079-3.

Santos, E., S. R. Tronick, S. A. Aaronson, S. Pulciani, and M. Barbacid. 1982. "T24 human bladder carcinoma oncogene is an activated form of the normal human homologue of BALB- and Harvey-MSV transforming genes." *Nature* 298(5872):343–7. doi: 10.1038/298343a0.

Sapranauskas, R., G. Gasiunas, C. Fremaux, R. Barrangou, P. Horvath, and V. Siksnys. 2011. "The Streptococcus thermophilus CRISPR/Cas system provides immunity in Escherichia coli." *Nucleic Acids Res* 39(21):9275–82. doi: 10.1093/nar/gkr606.

Satir, P., L. B. Pedersen, and S. T. Christensen. 2010. "The primary cilium at a glance." *J Cell Sci* 123(4):499–503. doi: 10.1242/jcs.050377.

Schlesinger, D. H., and G. Goldstein. 1975. "Molecular conservation of 74 amino acid sequence of ubiquitin between cattle and man." *Nature* 255(5507):423–4. doi: 10.1038/255423a0.

Scientia. 2017. "Interview with NIGMS Director Dr. Jon Lorsch." *Scientia* Aug. 24 2017.

Scolnick, E. M., A. G. Papageorge, and T. Y. Shih. 1979. "Guanine nucleotide-binding activity as an assay for src protein of rat-derived murine sarcoma viruses." *Proc Natl Acad Sci USA* 76(10):5355–9. doi: 10.1073/pnas.76.10.5355.

Scolnick, E. M., and W. P. Parks. 1974. "Harvey sarcoma virus: A second murine type C sarcoma virus with rat genetic information." *J Virol* 13(6):1211–9. doi: 10.1128/JVI.13.6.1211-1219.1974.

Scolnick, E. M., E. Rands, D. Williams, and W. P. Parks. 1973. "Studies on the nucleic acid sequences of Kirsten sarcoma virus: A model for formation of a mammalian RNA-containing sarcoma virus." *J Virol* 12(3):458–63. doi: 10.1128/JVI.12.3.458-463.1973.

Scotto, J., and J. C. Bailar, 3rd. 1969. "Rigoni-Stern and medical statistics. A nineteenth-century approach to cancer research." *J Hist Med Allied Sci* 24(1):65–75. doi: 10.1093/jhmas/xxiv.1.65.

Shah, S. P., A. Roth, R. Goya, A. Oloumi, G. Ha, Y. Zhao, G. Turashvili, J. Ding, K. Tse, G. Haffari, A. Bashashati, L. M. Prentice, J. Khattra, A. Burleigh, D. Yap, V. Bernard, A. McPherson, K. Shumansky, A. Crisan, R. Giuliany, A. Heravi-Moussavi, J. Rosner, D. Lai, I. Birol, R. Varhol, A. Tam, N. Dhalla, T. Zeng, K. Ma, S. K. Chan, M. Griffith, A. Moradian, S. W. Cheng, G. B. Morin, P. Watson, K. Gelmon, S. Chia, S. F. Chin, C. Curtis, O. M. Rueda, P. D. Pharoah, S. Damaraju, J. Mackey, K. Hoon, T. Harkins, V. Tadigotla, M. Sigaroudinia, P. Gascard, T. Tlsty, J. F. Costello, I. M. Meyer, C. J. Eaves, W. W. Wasserman, S. Jones, D. Huntsman, M. Hirst, C. Caldas, M. A. Marra, and S. Aparicio. 2012. "The clonal and mutational evolution spectrum of primary triple-negative breast cancers." *Nature* 486(7403):395–9. doi: 10.1038/nature10933.

Shamblott, M. J., J. Axelman, S. Wang, E. M. Bugg, J. W. Littlefield, P. J. Donovan, P. D. Blumenthal, G. R. Huggins, and J. D. Gearhart. 1998. "Derivation of pluripotent stem cells from cultured human primordial germ cells." *Proc Natl Acad Sci USA* 95(23):13726–31. doi: 10.1073/pnas.95.23.13726.

Shepard, H. M., G. L. Phillips, D. Thanos C, and M. Feldmann. 2017. "Developments in therapy with monoclonal antibodies and related proteins." *Clin Med (Lond)* 17(3):220–32. doi: 10.7861/clinmedicine.17-3-220.

Shibuya, M., S. Yamaguchi, A. Yamane, T. Ikeda, A. Tojo, H. Matsushime, and M. Sato. 1990. "Nucleotide sequence and expression of a novel human receptor-type tyrosine kinase gene (flt) closely related to the fms family." *Oncogene* 5(4):519–24.

Shih, C., B. Z. Shilo, M. P. Goldfarb, A. Dannenberg, and R. A. Weinberg. 1979. "Passage of phenotypes of chemically transformed cells via transfection of DNA and chromatin." *Proc Natl Acad Sci USA* 76(11):5714–8. doi: 10.1073/pnas.76.11.5714.

Shih, C., and R. A. Weinberg. 1982. "Isolation of a transforming sequence from a human bladder carcinoma cell line." *Cell* 29(1):161–9. doi: 10.1016/0092-8674(82)90100-3.

Shih, T. Y., M. O. Weeks, H. A. Young, and E. M. Scholnick. 1979. "Identification of a sarcoma virus-coded phosphoprotein in nonproducer cells transformed by Kirsten or Harvey murine sarcoma virus." *Virology* 96(1):64–79. doi: 10.1016/0042-6822(79)90173-9.

Shimizu, K., M. Goldfarb, M. Perucho, and M. Wigler. 1983. "Isolation and preliminary characterization of the transforming gene of a human neuroblastoma cell line." *Proc Natl Acad Sci USA* 80(2):383–7. doi: 10.1073/pnas.80.2.383.

Shimomura, O., F. H. Johnson, and Y. Saiga. 1962. "Extraction, purification and properties of aequorin, a bioluminescent protein from the luminous hydromedusan, Aequorea." *J Cell Comp Physiol* 59:223–39. doi: 10.1002/jcp.1030590302.

Siddiqui, M. Z. 2010. "Monoclonal antibodies as diagnostics; an appraisal." *Indian J Pharm Sci* 72(1):12–7. doi: 10.4103/0250-474X.62229.

Sigismund, S., D. Avanzato, and L. Lanzetti. 2018. "Emerging functions of the EGFR in cancer." *Mol Oncol* 12(1):3–20. doi: 10.1002/1878-0261.12155.

Silverstein, A. M. 2000. "The most elegant immunological experiment of the XIX century." *Nat Immunol* 1(2):93–4. doi: 10.1038/77874.

Silverstein, A. M. 2011. "Ilya Metchnikoff, the phagocytic theory, and how things often work in science." *J Leukoc Biol* 90(3):409–10. doi: 10.1189/jlb.0511234.

Sistonen, L., and K. Alitalo. 1986. "Activation of c-ras oncogenes by mutations and amplification." *Ann Clin Res* 18(5–6):297–303.

Sjostrand, F. S. 1953. "Electron microscopy of mitochondria and cytoplasmic double membranes." *Nature* 171(4340):30–2. doi: 10.1038/171030a0.

Solovyeva, V. V., A. A. Shaimardanova, D. S. Chulpanova, K. V. Kitaeva, L. Chakrabarti, and A. A. Rizvanov. 2018. "New approaches to Tay-Sachs disease therapy." *Front Physiol* 9:1663. doi: 10.3389/fphys.2018.01663.

Solter, D. 2006. "From teratocarcinomas to embryonic stem cells and beyond: A history of embryonic stem cell research." *Nat Rev Genet* 7(4):319–27. doi: 10.1038/nrg1827.

Sorokin, S. 1962. "Centrioles and the formation of rudimentary cilia by fibroblasts and smooth muscle cells." *J Cell Biol* 15:363–77. doi: 10.1083/jcb.15.2.363.

Soussi, T. 2010. "The history of p53. A perfect example of the drawbacks of scientific paradigms." *EMBO Rep* 11(11):822–6. doi: 10.1038/embor.2010.159.

Science News Staff, Science News. 2020. *Science*. doi: 10.1126/science.abc9393

Stehelin, D., H. E. Varmus, J. M. Bishop, and P. K. Vogt. 1976. "DNA related to the transforming gene(s) of avian sarcoma viruses is present in normal avian DNA." *Nature* 260 (5547):170–3. doi: 10.1038/260170a0.

Steinitz, M., G. Klein, S. Koskimies, and O. Makel. 1977. "EB virus-induced B lymphocyte cell lines producing specific antibody." *Nature* 269(5627):420–2. doi: 10.1038/269420a0.

Stephenson, J. A., J. C. Goddard, O. Al-Taan, A. R. Dennison, and B. Morgan. 2013. "Tumour Angiogenesis: A Growth Area—From John Hunter to Judah Folkman and beyond." *J Cancer Res* 2013:1–6. doi: 10.1155/2013/895019

Stern, A. M., and H. Markel. 2005. "The history of vaccines and immunization: Familiar patterns, new challenges." *Health Aff (Millwood)* 24(3):611–21. doi: 10.1377/hlthaff.24.3.611.

Sutton, W. S. 1903. "The chromosomes in heredity." *Biol Bull* 4(5):231–51.

Svedberg, T., and K. O. Pedersen. 1940. *The Ultracentrifuge*. Oxford, UK: Clarendon Press.

Szostak, J. W., and E. H. Blackburn. 1982. "Cloning yeast telomeres on linear plasmid vectors." *Cell* 29(1):245–55. doi: 10.1016/0092-8674(82)90109-x.

Takahashi, K., K. Tanabe, M. Ohnuki, M. Narita, T. Ichisaka, K. Tomoda, and S. Yamanaka. 2007. "Induction of pluripotent stem cells from adult human fibroblasts by defined factors." *Cell* 131(5):861–72. doi: 10.1016/j.cell.2007.11.019.

Takahashi, K., and S. Yamanaka. 2006. "Induction of pluripotent stem cells from mouse embryonic and adult fibroblast cultures by defined factors." *Cell* 126(4):663–76. doi: 10.1016/j.cell.2006.07.024.

Takahashi, T., M. M. Nau, I. Chiba, M. J. Birrer, R. K. Rosenberg, M. Vinocour, M. Levitt, H. Pass, A. F. Gazdar, and J. D. Minna. 1989. "p53: A frequent target for genetic abnormalities in lung cancer." *Science* 246(4929):491–4. doi: 10.1126/science.2554494.

Talmage, D. W. 1957. "Allergy and Immunology." *Annu Rev Med* 8:239–56. doi: 10.1146/annurev.me.08.020157.001323.

Tang, H., and J. B. Shrager. 2016. "CRISPR/Cas-mediated genome editing to treat EGFR-mutant lung cancer: A personalized molecular surgical therapy." *EMBO Mol Med* 8(2):83–5. doi: 10.15252/emmm.201506006.

Temin, H. M., and S. Mizutani. 1970. "RNA-dependent DNA polymerase in virions of Rous sarcoma virus." *Nature* 226(5252):1211–3. doi: 10.1038/2261211a0.

Temin, H. M., and H. Rubin. 1958. "Characteristics of an assay for Rous sarcoma virus and Rous sarcoma cells in tissue culture." *Virology* 6(3):669–88. doi: 10.1016/0042-6822(58)90114-4.

Terman, B. I., M. Dougher-Vermazen, M. E. Carrion, D. Dimitrov, D. C. Armellino, D. Gospodarowicz, and P. Bohlen. 1992. "Identification of the KDR tyrosine kinase as a receptor for vascular endothelial cell growth factor." *Biochem Biophys Res Commun* 187(3):1579–86. doi: 10.1016/0006-291x(92)90483-2.

Thiersch, K. 1865. *Der Epithelialkrebs, Namenthlich Der Haut Mit Atlas*. Leipzig, Germany: Wilhelm Engelmann.

Thomas, E. D. 2000. "Landmarks in the development of hematopoietic cell transplantation." *World J Surg* 24(7):815–8. doi: 10.1007/s002680010130.

Thomson, J. A., J. Itskovitz-Eldor, S. S. Shapiro, M. A. Waknitz, J. J. Swiergiel, V. S. Marshall, and J. M. Jones. 1998. "Embryonic stem cell lines derived from human blastocysts." *Science* 282(5391):1145–7. doi: 10.1126/science.282.5391.1145.

Tonegawa, S., C. Steinberg, S. Dube, and A. Bernardini. 1974. "Evidence for somatic generation of antibody diversity." *Proc Natl Acad Sci USA* 71(10):4027–31. doi: 10.1073/pnas.71.10.4027.

Tsien, R. Y. 1998. "The green fluorescent protein." *Annu Rev Biochem* 67:509–44. doi: 10.1146/annurev.biochem.67.1.509.

Underwood, E. A. 1943. Section of the history of medicine. In: *Lavoisier and the History of Respiration*. Vol. 37. Edited by W. Langdon-Brown. United Kingdom:Royal Society of Medicine.

Valent, P., B. Groner, U. Schumacher, G. Superti-Furga, M. Busslinger, R. Kralovics, C. Zielinski, J. M. Penninger, D. Kerjaschki, G. Stingl, J. S. Smolen, R. Valenta, H. Lassmann, H. Kovar, U. Jager, G. Kornek, M. Muller, and F. Sorgel. 2016. "Paul Ehrlich (1854–1915) and His contributions to the foundation and birth of translational medicine." *J Innate Immun* 8(2):111–20. doi: 10.1159/000443526.

van der Weyden, C. A., S. A. Pileri, A. L. Feldman, J. Whisstock, and H. M. Prince. 2017. "Understanding CD30 biology and therapeutic targeting: A historical perspective providing insight into future directions." *Blood Cancer J* 7(9):e603. doi: 10.1038/bcj.2017.85.

Virchow, R. 1863. Die krankhaften Geschwulste. *Hirschwald*.

Vogt, P. K. 2010. "Oncogenes and the revolution in cancer research: Homage to hidesaburo hanafusa (1929–2009)." *Genes Cancer* 1(1):6–11. doi: 10.1177/1947601909356102.

von Behring, E., and S. Kitasato. 1890. "[The mechanism of diphtheria immunity and tetanus immunity in animals. 1890]." *Mol Immunol* 28(12):1317, 1319–20.

Wainwright, M. 1987. "The history of the therapeutic use of crude penicillin." *Med Hist* 31(1):41–50.

Wainwright, M. 1989. "Moulds in ancient and more recent medicine." *Mycology* 3(1):21–33.

Wang, L. H., P. Duesberg, K. Beemon, and P. K. Vogt. 1975. "Mapping RNase T1-resistant oligonucleotides of avian tumor virus RNAs: Sarcoma-specific oligonucleotides are near the poly(A) end and oligonucleotides common to sarcoma and transformation-defective viruses are at the poly(A) end." *J Virol* 16(4):1051–70. doi: 10.1128/JVI.16.4.1051-1070.1975.

Wang, W., G. Karamanlidis, and R. Tian. 2016. "Novel targets for mitochondrial medicine." *Sci Transl Med* 8(326):326rv3. doi: 10.1126/scitranslmed.aac7410.

Warburg, O. 1913. "Uber Sauerstoffatmende Koernchen aus Leberzellen und uber Sauerstoffatmung in Berekfeld-Filtralen Wassriger Leverextrakte." *Arch Gesamte Physiol* 154(9–10):599–617.

Watson, J. D. 1972. "Origin of concatemeric T7 DNA." *Nat New Biol* 239(94):197–201. doi: 10.1038/newbio239197a0.

Watson, J. D., and F. H. Crick. 1953. "Genetical implications of the structure of deoxyribonucleic acid." *Nature* 171(4361):964–7. doi: 10.1038/171964b0.

Waxman, L., J. M. Fagan, and A. L. Goldberg. 1987. "Demonstration of two distinct high molecular weight proteases in rabbit reticulocytes, one of which degrades ubiquitin conjugates." *J Biol Chem* 262(6):2451–7.

Weintraub, P. 2010. "The doctor who drank infectious broth, gave himself an ulcer, and solved a medical mystery." *Discovery Magazine*.

Weiss, R. A. 2020. "A perspective on the early days of RAS research." *Cancer Metastasis Rev* 39(4):1023–8. doi: 10.1007/s10555-020-09919-1.

Wennergren, G., and H. Lagercrantz. 2007. ""One sometimes finds what one is not looking for" (Sir Alexander Fleming): the most important medical discovery of the 20th century." *Acta Paediatr* 96(1):141–4. doi: 10.1111/j.1651-2227.2007.00098.x.

Wigler, M., A. Pellicer, S. Silverstein, and R. Axel. 1978. "Biochemical transfer of single-copy eucaryotic genes using total cellular DNA as donor." *Cell* 14(3):725–31. doi: 10.1016/0092-8674(78)90254-4.

Wilkinson, K. D., M. K. Urban, and A. L. Haas. 1980. "Ubiquitin is the ATP-dependent proteolysis factor I of rabbit reticulocytes." *J Biol Chem* 255(16):7529–32.

Wilson, E. B. 1907. "Supernumerary Chromosomes in Hemiptera." *Science* 26(677):870–1.

Wolpert, L. 1995. "Evolution of the cell theory." *Philos Trans R Soc Lond B* 349(1329):227–33. doi: 10.1098/rstb.1995.0106.

Xu, Y., and A. Goldkorn. 2016. "Telomere and Telomerase Therapeutics in cancer." *Genes (Basel)* 7(6):22. doi: 10.3390/genes7060022.

Yachida, S., C. M. White, Y. Naito, Y. Zhong, J. A. Brosnan, A. M. Macgregor-Das, R. A. Morgan, T. Saunders, D. A. Laheru, J. M. Herman, R. H. Hruban, A. P. Klein, S. Jones, V. Velculescu, C. L. Wolfgang, and C. A. Iacobuzio-Donahue. 2012. "Clinical significance of the genetic landscape of pancreatic cancer and implications for identification of potential long-term survivors." *Clin Cancer Res* 18(22):6339–47. doi: 10.1158/1078-0432.CCR-12-1215.

Youn, Y. H., and Y. G. Han. 2018. "Primary cilia in brain development and diseases." *Am J Pathol* 188(1):11–22. doi: 10.1016/j.ajpath.2017.08.031.

Yu, J., M. A. Vodyanik, K. Smuga-Otto, J. Antosiewicz-Bourget, J. L. Frane, S. Tian, J. Nie, G. A. Jonsdottir, V. Ruotti, R. Stewart, I. I. Slukvin, and J. A. Thomson. 2007. "Induced pluripotent stem cell lines derived from human somatic cells." *Science* 318(5858):1917–20. doi: 10.1126/science.1151526.

Zakrzewski, W., M. Dobrzynski, M. Szymonowicz, and Z. Rybak. 2019. "Stem cells: past, present, and future." *Stem Cell Res Ther* 10(1):68. doi: 10.1186/s13287-019-1165-5.

Zimmermann, K. W. 1898. "Beitrage zur Kenntniss einiger Drusen und Epithelien." *Arch mikrosk Anat* 52(3):552–706.

Index

Note: Locators in *italics* represent figures and **bold** indicate tables in the text.

9781032065083